A Beginner's Guide to Biotechnology

quantum scientific publishing

A Beginner's Guide to Biotechnology

BioPharmaceutical Technology Center Institute

quantum scientific publishing

A Beginner's Guide to Biotechnology

ISBN-13: 978-1481186131
ISBN-10: 1481186132

Published by quantum scientific publishing

Pittsburgh, PA | Copyright © 2012

Cover design by Scott Sheariss

QUANTUM
SCIENTIFIC
PUBLISHING

Table of Contents

Unit One

Unit Two

Unit Three

Appendix

Unit One

Section 1.1 – The History of Biotechnology

Section Objectives

- Discuss the historical context in which the field of biotechnology emerged by identifying key scientific developments

- Explain molecular biology and its role in biotechnology

What is Biotechnology?

Biotechnology is a blended field of science and industry that includes the study of organisms and their molecular building blocks (DNA, RNA, proteins, and carbohydrates) and how these molecules might be used to benefit different areas, such as food production and **fermentation** (beer, wine, and cheese to name only a few), farming, **pharmaceuticals,** environmental cleanup, energy production, and almost any other field in which living things are a necessary part. In fact, it is the interdisciplinary nature of biotechnology and its vast potential that makes it so exciting! The difference between basic scientific research into organisms (their molecular components) and biotechnology is the applied nature of biotechnology, such that these organisms or biological components are utilized to make a product or process for human use.

Biotechnology has a rich history. In the 1850s, Gregor Mendel was the first person to trace the inheritance patterns of the characteristics of pea plants from parent to offspring. Thus, he is credited with being the "father of genetics." Examples of characteristics that he studied are seed color (yellow or green), seed texture (rough or smooth), and plant height (short or tall). Mendel noted that some traits appeared more often, or dominated, the alternative, or recessive, form. This concept is now known as Mendelian genetics. In Mendel's time, scientists did not know about DNA. They could track the traits or phenotypes, of organisms, but they did not know the molecules that were involved in the process.

Biotechnology

The use of living organisms and their sub-parts for the creation of commercial products such as genetically modified crops and pharmaceuticals.

Fermentation

Growth of bacteria or yeast in a large culture vessel under tightly controlled conditions that maximize cell growth and DNA or protein production; often used to modify one energy source to produce a desired byproduct (for example ethanol production in beer or wine).

Pharmaceuticals

Drugs that are useful for various purposes in healthcare.

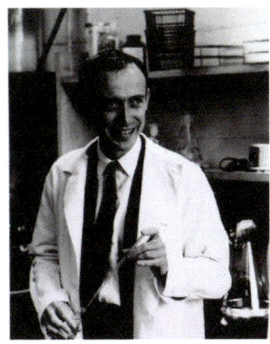
James Watson

Maurice Wilkins

Francis Crick

Rosalind Franklin

Alfred Hershey

Famous scientists who played key roles in the founding of biotechnology

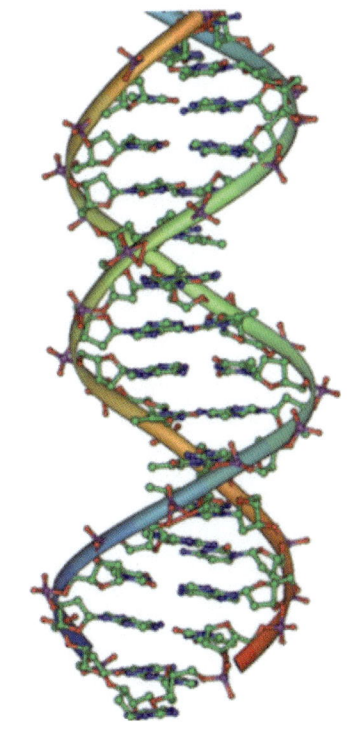
Double helical chemical structure of DNA

Deoxyribonucleic Acid (DNA)

The chemical constituent of the nucleus that makes up genes – the molecular basis of heredity; DNA consist of a double helix and has a sugar-phosphate backbone with purine (adenine, guanine) and pyrimidine (cytosine, thymine) bases; the sequence of these purines and pyrimidines is what makes up a particular gene.

DNA and Biotechnology

Let's start with **DNA**. DNA can be thought of as the molecule that stores the genetic information in a cell and organism. It is the blueprint for all that goes on within them. DNA is found in the nucleus of the cells and is organized into genes. Information that is needed for cellular processes is passed via the molecule ribonucleic acid (RNA) in a process called transcription. There are many kinds of RNA in the cells, but we are especially interested in messenger RNA, or mRNA, which is a copy of the DNA gene. mRNA is central to information flow within the cell because, although it starts in the nucleus, it moves into the cytoplasm, where it is translated into a protein in a process called translation.

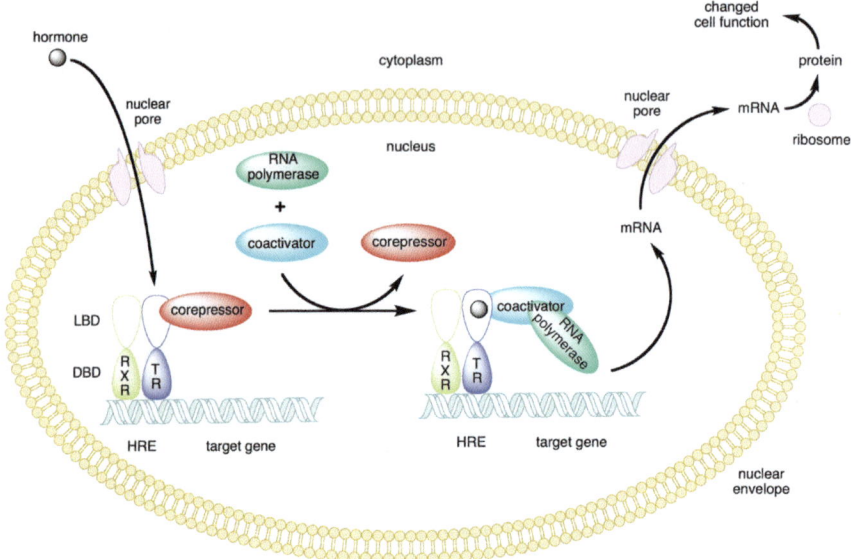

RNA transcription and protein translation

Proteins and Biotechnology

Each gene contains the information to make a different protein. These different proteins are useful for many different purposes. This flow of information from the genes is known as the **central dogma of molecular biology** (DNA to RNA to proteins).

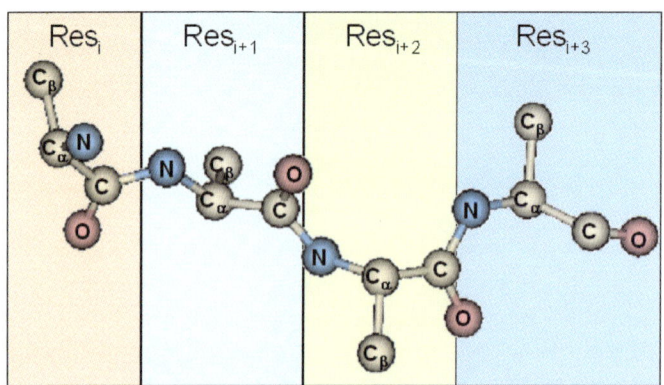

Polypeptide chain composed of individual amino acids.

12

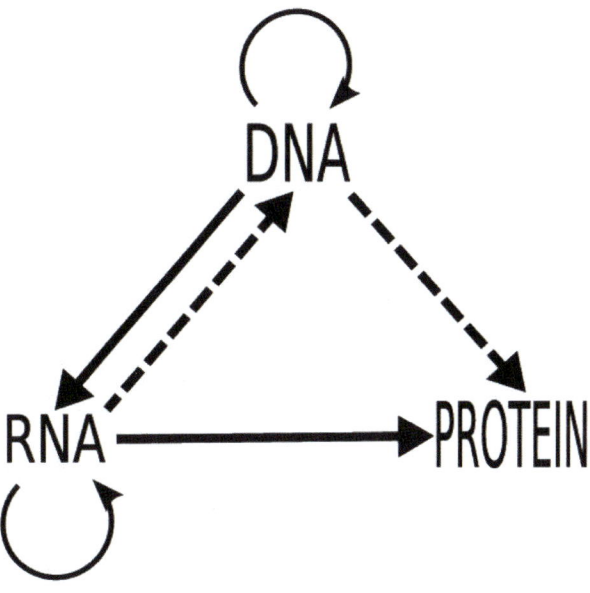

Central dogma of molecular biology

As an example, the protein insulin is important for the proper processing of the sugars in the foods that we eat. In order to produce insulin, our pancreas cells use the DNA code for insulin – the insulin gene – and create an mRNA copy. The mRNA copy then moves to the cytoplasm, where it is translated into the insulin protein. The insulin protein is then released into the bloodstream so that our bodies are better able to utilize the sugar from the foods we eat to provide energy for our cells. People with diabetes often do not produce enough insulin naturally to allow them to process sugar, so biotechnology has been utilized to allow bacteria to produce human insulin for use in diabetic patients. This is a much less expensive and safer source of insulin, as compared to the older versions, which were purified from cow or pig sources or from human cadavers. Thus, biotechnology was able to engineer bacteria to be "insulin factories."

> **Central Dogma of Molecular Biology**
>
> DNA is the genetic code for RNA, which is then translated into protein.

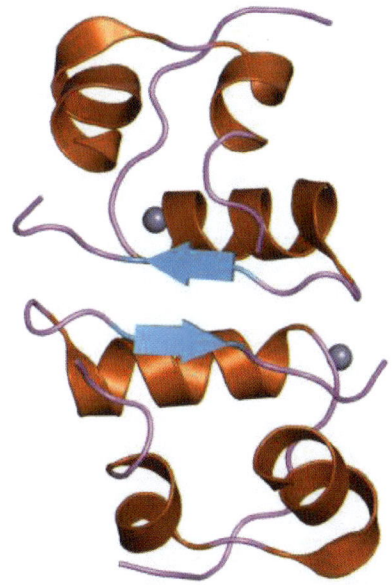

Structure of insulin protein

There are many, many proteins that make up the wide variety of organisms inhabiting the Earth. Thus, it is useful to classify these proteins into categories based on their biological functions (enzymes, antibodies, hormones, etc.). The scientists George Beadle and Edward Tatum are credited with the discovery that one gene codes for one protein. This discovery made an important link between genetics and **proteomics,** since we depend on genes and gene products for almost everything we do in the field of biotechnology! In humans, there are around 25,000-30,000 genes.

It is now known that each gene can actually code for more than one protein. This is because the genes can create variations of themselves during transcription. This is called splicing. Each of these variants serves as the code for additional unique proteins. In addition, proteins can be decorated with other biomolecules (sugars and fats to name a few), such that there are millions of proteins that can be made by any cell or organism. Researchers are constantly discovering new proteins and studying their functions, some of which become useful in biotechnology.

Edward Tatum

George Beadle

Molecular Biology and Biotechnology

Why is knowledge of molecular biology so important for the field of biotechnology? The simple answer is that basic molecular biology principles are the foundation on which the key tools for the biotechnology industry are based. These tools are then used to create and harness organisms and molecular components for applications in agriculture, pharmaceuticals, energy production, environmental cleanup, and other fields.

All of these applications involve the use of molecular biology tools. What began as basic research on the workings of the cell and of organisms has become the knowledge that has allowed us to extend our abilities in other fields. The interdisciplinary use of molecular biology tools through biotechnology is what makes biotechnology so interesting and exciting, providing unlimited potential for biotechnology in the future. Many current problems in the world may someday be solved through the use of biotechnology.

Summary

Biotechnology involves the use of biological molecules or systems for use in food production, pharmaceutical development, and industrial applications to name only a few. It has a rich history and involves the important discoveries of numerous scientists around the world. Most critical was the discovery of DNA and its function as the hereditary molecule in each cell. The manipulation of DNA in molecular biology has become the foundation of modern biotechnology.

Concept Reinforcement

1. What are some ways in which biotechnology has influenced the world around you?

2. What are some other phenotypes, or traits, of various organisms?

3. What is the molecule in each cell that is responsible for heredity?

Section 1.2 – Advances in Biotechnology

Section Objectives

- Describe classical and modern biotechnology

- Describe molecular biology and its relationship to biotechnology

- Discuss the connections between basic biotechnology tools and their applications

Classical versus Modern Biotechnology

Biotechnology is the use of living organisms, biological processes, or their component parts to solve problems or make useful commercial products. It involves modifying or manipulating organisms, often at the molecular level, to create new and practical applications for agriculture, medicine, and industry.

Humans have been practicing **traditional or classical biotechnology** for thousands of years through the domestication of dogs, cattle, and other animals, the selective breeding of plants for desired characteristics, and the production of various food products, such as cheese, bread, beer, yogurt, and wine through microbial **fermentation.** Humans have been manipulating their environment for millennia to improve their way of life.

Modern biotechnology arose in the last 150 years, due to the successive discoveries of the structures and functions of the key biological molecules in the cell or organism (DNA, RNA, and proteins) by numerous scientists throughout the world. These key researchers include, to name just a few, Gregor Mendel in Austria, Louis Pasteur in France, Watson and Crick in England, and Linus Pauling, Arthur Kornberg, Paul Berg, Francis Collins, and Craig Venter in the United States. James Watson, Francis Collins, and Craig Venter were instrumental in the completion of the **Human Genome Project** in 2003, which identified the complete sequence of human DNA in all 23 chromosomes. This, in turn, has allowed for the identification of ~30,000 human genes, most of whose functions are still being investigated. In addition to the humane genome, the complete genomes of over 100 different organisms have now been sequenced, including the chimpanzee, protozoa and ciliates, numerous bacteria and yeast, the fruit fly, mosquitoes, honey bees, nematode worms, the mouse, the rat, algae, rice, the rabbit, the cow, the dog, the cat, and the chicken. Many of these organisms are considered model organisms for research. They may also cause disease in humans or animals and are thus important for medical and/or veterinary research. This genetic sequence information will also advance biotechnology as novel uses for unique gene products are found and applied in a wide variety of organisms.

Traditional or Classical Biotechnology

Use of or modification of plants, animals, or microbes for the production of foods and other products; first occurred at the beginning of human civilization.

Fermentation

Growth of bacteria or yeast in a large culture vessel under tightly controlled conditions that maximize cell growth and DNA or protein production; often used to modify one energy source to produce a desired by-product (for example ethanol production in beer or wine).

Modern Biotechnology

Use of or modification of organisms and their parts at the molecular level for the production of products that effect human health or quality of life..

Arthur Kornberg

Paul Berg

Louis Pasteur

Francis Collins

Craig Venter

Craig Venter image courtesy of:
http://biology.plosjournals.org/perlserv/?request=slideshow&type=figure&doi=10.1371/journal.pbio.0050266&id=85043

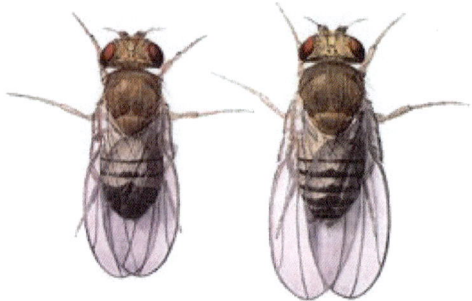

Fruit Fly *(Drosophila melanogaster)*. Male on left. Female on right.
Image courtesy of NASA

18

Chimpanzee (*Pan troglodytes*)

Rice
Image courtesy of the US Department of Agriculture

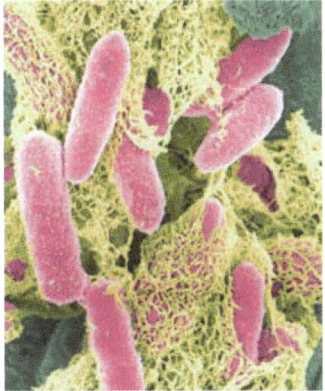

Bacteria (*Escherichia coli*)
Image courtesy of the US Government

Molecular Biology

The study of biology at the molecular level to understand the various interactions and functions of the molecules in a cell or organism, particularly DNA, RNA, proteins, lipids (fats), and carbohydrates (sugars).

Recombinant DNA (*r*DNA)

Combining (joining) DNA sequences or pieces that do not normally occur together.

GMO

Genetically modified organism; a living thing that has had its DNA altered, usually with the addition of genetic material from another organism that confers a desirable trait.

Examples of discoveries and applications in biotechnology

Traditional (Classical) Biotechnology	Modern Biotechnology
Animal breeding	Genetically modified animals (GMOs)
Plant breeding	Genetically modified plants (GMOs)
Microbial fermentation (bread, cheese, yogurt, wine, and beer)	Microbial expression of proteins and other products, such as antibiotics or industrial enzymes
	Human medicines, such as antibodies, vaccines, and gene therapy
	Human medical diagnostic devices
	Environmental cleanup using microbes for toxic waste, oil spills, and human and animal waste products
	Bioenergy production using plants or microbes

Molecular Biology and Biotechnology

The term "**molecular biology**" was coined in 1938 by Warren Weaver, who was at the time the director of natural sciences for the Rockefeller Foundation. Molecular biology is the basis for modern biotechnology, with an emphasis on specific applications. While the tools needed for biotechnology vary depending on the application and may be widely used or specific to a segment of biotechnology, **recombinant DNA** technology has found numerous applications in biotechnology. It allows scientists to combine various DNA genes from different organisms, such as bacteria, yeast, plants, and animals. Many of the applications of recombinant DNA technology are listed in the table. These applications include genetically modified organisms (**GMOs**), proteins (antibodies and enzymes), and human and animal vaccines. In addition, genetically engineered bacteria and other microbes may become widely used to extract oil and valuable metals from the ground and industrial waste sites (biomining and bioremediation), while algae are being investigated as possible sources of biologically generated fuels, such as biodiesel (bioenergy).

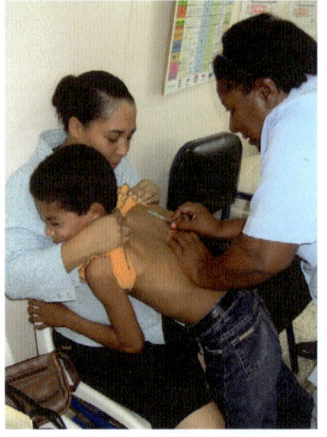

Vaccination

20

Genetically modified soybean field

Glyphosate treatment of regular plant versus plant engineered to be resistant
to an herbicide (i.e. RoundUp Ready plant versus unmodified plant)

Summary

Traditional biotechnology has been around for thousands of years and involved primarily the use of microorganisms to produce fermented food or beverages. Modern biotechnology has evolved since the advent of molecular biology and the capability to purify and study the various biological molecules in cells, and thus take advantage of their unique properties. The manipulation of DNA has allowed for the generation of genetically modified organisms and the production of industrial and pharmaceutical proteins. In addition, biotechnology is expanding in the areas of biofuels, bioremediation, and biomining.

Concept Reinforcement

1. What are the differences between traditional biotechnology and modern biotechnology?

2. What is molecular biology?

3. What is the relationship between molecular biology and biotechnology?

Section 1.3 – Global and Interdisciplinary Nature of Biotechnology

Section Objectives

- Describe the ways in which biotechnology is a global industry

- Discuss the interdisciplinary nature of biotechnology

- Describe bioinformatics and its role in biotechnology

Biotechnology as an Interdisciplinary Field

Biotechnology combines business, science, legal, and manufacturing expertise to produce products or services that are useful to civilization. Although each of these areas is important to business, biotechnology is one of the most research-intensive industries in the world. Today's biotechnology research has its roots in not only molecular and cellular biology, but many other disciplines as well, such as chemistry, physics, mathematics, engineering, and computer science. There is a strong interplay of the life sciences with other sciences (i.e. math, physics and engineering) and this interplay is crucial for many advances in biotechnology. For example, a single biotechnology product, such as an antibody drug, may involve discoveries not only in biology, but also chemistry to provide the proper formulation of the drug, engineering to provide the proper delivery device for the drug and proper manufacturing equipment, and computer science to generate the software to analyze and monitor the information from the human clinical trials for the drug.

As biotechnology progresses, the importance of engineering and computer science are becoming more and more critical. The new field of **bioinformatics** combines the use of computer science to develop computer programs and algorithms to capture, organize, and analyze vast amounts of complex scientific data that would be impossible to analyze manually. One application of bioinformatics is to allow scientists to determine the three-dimensional structure of biological molecules, such as proteins, so they can readily study them and predict their function. In addition, programs to study, analyze, and compare DNA and RNA sequences are also available.

> **Bioinformatics**
> Combination of mathematics, computer science algorithms, and often statistics to answer complex biological problems, usually through data analysis.

Molecular biotechnology

Bioinformatics

The field of bioinformatics has been instrumental to the development and advancement of biotechnology, which allows for the analysis of the vast amounts of biological information being generated. Advances in biotechnology occur not only in the techniques used to analyze and produce various biological molecules, but also the computer software required to analyze the results. Bioinformatics focuses on the analysis of gene and protein sequences and structures, in addition to complicated data analysis. Resources have been developed that allow for determining how similar the DNA sequences of two or more genes are or how a protein may fold and exhibit a certain function. They can quickly determine relatedness and thus are instrumental in tracking evolution. Such databases are generally publicly accessible and free of charge. For example, a DNA database called "Genbank" is available on the internet, as is a protein database called "SwissProt." Bioinformatics is also being used to a analyze gene expression of many different mRNAs at once (thousands), the analysis of mutations involved in cancer and other diseases, and image analysis of complex cells. Bioinformatics will continue to be a very important component of biotechnology.

GenBank, NIH's publicly accessible genetic sequence database, was formed at Los Alamos National Laboratory. Scientists submit DNA sequence data from a wide range of organisms to GenBank. Researchers routinely retrieve and analyze the data in the archive.Image courtesy of the National Human Genome Research Institute

Global Biotechnology

Just as biotechnology is interdisciplinary in nature, it is also global in nature. The development of modern biotechnology was the combined efforts of scientists throughout the world, and continues to be so today. Collaborations between scientists and organizations in different countries help drive biotechnology and also spread its availability to help everyone. Biotechnology is striving to find unique solutions for the world's problems that fit each local culture and environment.

As an example, over 56 different genetically modified crops have been tested in over 34 different countries, highlighting the global nature of biotechnology. Currently four countries (the United States, Argentina, Canada and China) produce 99% of the world's GMO crops. Other countries also produce GMO crops, but to a much lesser extent (Australia, South Africa, Mexico, Spain, France, Bulgaria, Romania, Uruguay, and Germany). Soybeans continue to be the most genetically modified crop, accounting for over 50% of the global biotechnology acreage. Most genetic modifications to crops increase resistance of the modified crop to commercially available herbicides (such as **RoundUp**) or introduce a foreign gene that allows the crop to be resistant to insect damage (such as **Bt**). However, other modifications include taste enhancement, increased storage properties, and increased nutritional content. As alternatives to either finding them in nature or producing them with harsh and expensive chemical processes, researchers are also investigating ways to use genetically modified plants to produce useful dyes, flavorings, drugs, and chemicals.

Glyphosate treatment of regular plant versus plant engineered to be resistant to an herbicide (i.e. RoundUp Ready plant versus unmodified plant.

In short, the desired outcome of the research on genetically modified crops will be to increase their texture, quality, variety, and availablity. However, nations throughout the world have had different responses to biotechnology, with different levels of consumer acceptance and government regulation. While the United States has generally had high levels of public confidence and market acceptance, the **European Union** has adopted stricter regulation because of more public criticism and concern.

Bt

Bacillus Thuringiensis. The Cry genes of this species of bacteria produce toxins that are harmful to the larvae of various moths, butterflies, mosquitoes, and beetles; when expressed in plants, it inhibits these insects from feeding on the modified plant; the bacterium itself is also used as a natural, organic, insecticide.

In the medical arena, continued effort in global biotechnology will be placed on disease diagnosis, prevention, and treatments. **Vaccines** for such diseases as **AIDS**, **malaria**, and **SARS** will dramatically increase human health throughout the world, but particularly in developing nations. Malaria is one of the most common infectious diseases and primarily affects people living in tropical and subtropical parts of the world and is a major hindrance to economic development in these regions. Thus there is continued effort worldwide to develop effective preventatives for malaria. Significant time and resources were used to produce a vaccine to animal **rabies**, which was initially developed in 1885 and then updated in 1967. This vaccine has saved countless human and animal lives.

Disease Name	Description
AIDS: Acquired Immune Deficiency Syndrome	Groups of symptoms and secondary diseases caused by infection with the human immunodeficiency virus (HIV); the disease has killed more than 25 million people to date and continues to spread throughout the world, though is most prevalent in Africa; antiretroviral therapies are available, but are often too expensive or unavailable for those areas.
Malaria	*Plasmodium* parasite. The parasite is contracted when bitten by an infected mosquito which allows the parasite to multiply in red bood cells. It kills between 1-3 million people world wide every year, with most of the deaths occurring in Africa. Preventative treatments are available, but are often too expensive or not available to those most affected.
Rabies	A viral (*Lyssavirus* genus) neuroinvasive disease that causes inflammation of the brain in mammals. It is most commonly spread by a bite from an infected animal and if left untreated is fatal. Rabies is found worldwide and causes about ~55,000 deaths per year.
SARS: Severe Acute Respiratory Syndrome	A respiratory disease caused by the SARS corona virus that can be particularly deadly in the elderly and immune-compromised people.

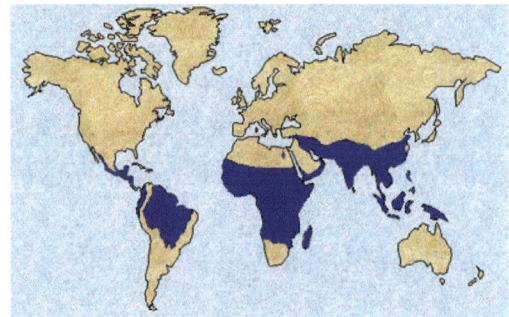

Map of malaria distribution throughout the world

In addition to vaccines, the development and purification of human antibodies to treat numerous diseases will continue in the future and throughout the world. There are currently at least 25 antibodies on the market to treat such diseases as asthma, rheumatoid arthritis, and cancer. This number will only increase, as current and new biotechnology companies devote resources to identifying new therapeutic antibodies or proteins. Many of the antibody products are from biotechnology companies. Genentech is considered the first biotechnology drug discovery company and successfully marketed the first genetically engineered drug in 1982, a form of human insulin produced in bacteria. Amgen is considered to be the largest biotechnology company in the world to date.

Examples of Therapeutic Antibodies Generated by Biotechnology

Brand Name of Drug	Company Producing It	Disease Targeted
Xolair	Genentech	Asthma
Humira	Abbott	Rheumatoid arthritis
Lucentis	Genentech	Macular degeneration
Erbitux	Imclone/BMS	Colorectal cancer
Remicade	Johnson and Johnson	Rheumatoid arthritis and Crohn's disease
Herceptin	Genentech	Metastatic breast cancer

Genentech company logo

Summary

Biotechnology is an industry that focuses heavily on science and is interdisciplinary in nature, since it involves the participation of those expert in math, physics, engineering, biology, chemistry, and computer science. Biotechnology is also very global in nature and affects people who live in all parts of the world. The development of genetically modified plants to aid in food production and of vaccines to prevent infectious disease are representative of the ways it is having an important impact on society now and will continue to do so in the future.

Concept Reinforcement

1. How do you think that math, chemistry, physics, engineering, and computer science might be important in biotechnology?

2. What does it mean for something we eat to be genetically modified?

3. What are some other characteristics that might be beneficial in food crops or other plants that biotechnology could produce?

4. How is the biotechnology industry global and how has it affected people in various parts of the world?

Section 1.4 – Intellectual Property, Regulation, and Quality in Biotechnology

Section Objectives

- Explain the role of patents and licensing processes in new biotechnology innovations

- Discuss the importance of quality in biotechnology

- Explain the regulations involved in biotechnology

- Describe the regulatory processes involved in the use of recombinant DNA technology

Intellectual Property

Biotechnology worldwide is driven by innovation and scientific discoveries, which can then be applied to real world problems. In the life sciences, the discovery may be of a novel enzyme that can perform an activity useful for an industrial process, it might be a technique that can be used to produce a new drug, or it may be a gene that can enhance the resistance of various crops to insect damage. Many of these new discoveries occur in academic laboratories in college and universities around the globe. The challenge for a biotechnology company is to use those basic discoveries and find applications for them that allow them to be useful and practical in our lives.

Research done in an industrial setting is highly controlled. Discoveries, whether they come from an academic laboratory (basic research lab, typically at a university) or an industrial laboratory (applied research lab, typically in a company), are almost always protected by a patent. A **patent** is legal protection for an invention that is granted by the US Patent and Trademark Office (USPTO) or an equivalent in another country. Each country has its own set of patent rules and regulations. The set of novel and unique information from scientific discoveries that is owned by a company is called its **intellectual property** (IP). Intellectual property is key to the success of a biotechnology company because it allows the owner to exclude (prevent) others from using the information. Biotechnology companies often will acquire patents from universities or other companies through **licensing** agreements, which allow the discovery to be developed by the company that obtains the license. The license usually gives financial rewards to the inventor, which is an incentive for the inventor to license patented technologies.

Patent

Set of exclusive rights granted to an inventor by a government for a fixed period of time in exchange for disclosing (describing in detail) the invention. The period in the US is currently 20 years. The claims in the patent must be inventive and useful for the patent to issue. The exclusive right granted allows the inventor to exclude others form using, making, selling, or importing the invention, but does not necessarily give them the right to practice the invention.

Intellectual Property (IP)

The patent portfolio held by an inventor or company that allows them the exclusive rights to the discoveries.

Licensing

An agreement between parties granting the licensee the use of the information or other intellectual property from the licensor. The information is usually held within a patent or group of patents.

X 000001
July 31, 1790

The United States.

To all to whom these Presents shall come, Greeting.

Whereas Samuel Hopkins of the city of Philadelphia and State of Pensylvania hath discovered an Improvement, not known or used before, such Discovery, in the making of Pot ash and Pearl ash by a new Apparatus and Process, that is to say, in the making of Pearl ash 1st. by burning the raw Ashes in a Furnace, 2d. by dissolving and boiling them when so burnt in Water, 3d. by drawing off and settling the ley, and 4th. by boiling the ley into Salts which then are the true Pearl ash, and also in the making of Pot ash so made as aforesaid; which Operation of burning the raw Ashes in a Furnace, preparatory to their Dissolution and boiling in Water, is new, leaves little Residuum; and produces a much greater Quantity of Salt: These are therefore in pursuance of the Act, entituled "An Act to promote the Progress of useful Arts", to grant to the said Samuel Hopkins, his Heirs, Administrators and Assigns, for the Term of fourteen Years, the sole and exclusive Right and Liberty of using and vending to others the said Discovery, of burning the raw Ashes previous to their being dissolved and boiled in Water, according to the true Intent and Meaning of the Act aforesaid. In Testimony whereof I have caused these Letters to be made patent, and the seal of the United States to be hereunto affixed. Given under my Hand at the City of New York this thirty first Day of July in the Year of our Lord one thousand seven hundred & Ninety.

G. Washington

City of New York July 31st. 1790.

I do hereby certify that this foregoing Letters patent were delivered to me in pursuance of the Act, entituled "An Act to promote the Progress of useful Arts", that I have examined the same, and find them conformable to the said Act.

Edm: Randolph *Attorney General for the United States.*

The first US patent issued in 1790

Regulation in Biotechnology

Research performed in an industrial setting is regulated by many different agencies, commissions, guidelines, and laws. Biotechnology research is usually done using good laboratory practice (GLP) guidelines. Any information documented in a laboratory notebook is considered legally binding and must follow a specific form when entered. The information present in a laboratory notebook can be used as supporting information for patent submissions. The US Food and Drug Administration (FDA) has rules for good laboratory practices that cover the safety (preclinical) trials performed on animals prior to human testing. Other government agencies may play a role in biotechnology research, as well. These include the Environmental Protection Agency (EPA) and the United States Department of Agriculture (USDA) if the biotechnology product or service is agricultural or will impose a risk to the environment.

Food and Drug Administration (FDA) logo

With the discovery of the processes necessary to manipulate DNA, and with the advent of modern recombinant DNA technology, various ethical issues arose surrounding its use. Following the publication of methods to allow the expression of foreign DNA genes in bacteria through recombinant DNA methods by Stanley Cohen and Herbert Boyer in 1974, an international meeting was held at Asilomar, California. At this meeting, scientists urged the government to adopt guidelines regulating recombinant DNA experimentation. The scientists recommended the development of safe bacteria and **plasmids** that would be used strictly for experimental research and could not escape laboratory use to harm people or the environment. The Recombinant DNA Advisory Committee (RAC) was then formed to monitor recombinant DNA technology and provide guidelines for its uses. Similar guidelines have also been recommended and established for the use of recombinant DNA in **gene therapy** trials on humans. The Recombinant DNA Advisory Committee is part of the Office of Biotechnology Activities of the National Institutes of Health (NIH).

In contrast to the GLP guidelines for regulating laboratory in a research environment, Good Manufacturing Practices (GMP) govern the work that goes on in a manufacturing laboratory. Each manufacturing site of a biotechnology company will generally institute many quality systems and procedures to ensure the quality of the products that it is producing. GMP refers to the set of regulations by the FDA that require manufacturers of certain types of products (including drugs, medical devices, some foods, and blood products), to take steps to ensure the safety, purity, and efficacy of the products they make. GMP requires that the manufacturing process is highly documented. The purpose of the documentation is to ensure quality, but also to minimize or eliminate instances of contamination or errors. It involves record keeping, personnel qualifications, sanitation, cleanliness, equipment upkeep, process validation, and complaint handling. Training is a large component of GMP. It is very important to make sure that all employees involved in a manufacturing process have the necessary expertise to perform their duties. Most GMP guidelines are very general so that companies can decide which procedures and controls are the best ones for them to use. This system may also be called "cGMP" for current good manufacturing practices. cGMP is applied to production of medicines for distribution to those who need them. Another example is vaccine development.

In addition to GMP regulations, many companies also follow guidelines provided by the International Organization of Standardization (ISO). ISO is a network of national standards institutes in 157 member countries. The headquarters of ISO is in Geneva, Switzerland, and acts to coordinate the network. Some members of the network are part of governmental organizations, while others are in the public sector. Companies can become certified in various levels of ISO, depending on the degree of adoption of the code of ethics, technical regulations, and intellectual rights policies. The standards regulated by ISO are very similar to GMP. All of the processes in the manufacturing process must be documented and all employees must be well trained. However, ISO tends to cover all aspects of how a company operates, and it not restricted to manufacturing.

Plasmid

Double-stranded, circular piece of DNA that is found in bacteria and yeast and is separate from the genomic or chromosomal DNA; often possess genes advantageous for different environmental conditions and can be passed from one organism to another.

Gene therapy

Insertion of a gene into a person's cells to treat a disease, usually replacing a protein that is either absent or abnormal through hereditary defects (i.e. DNA mutations).

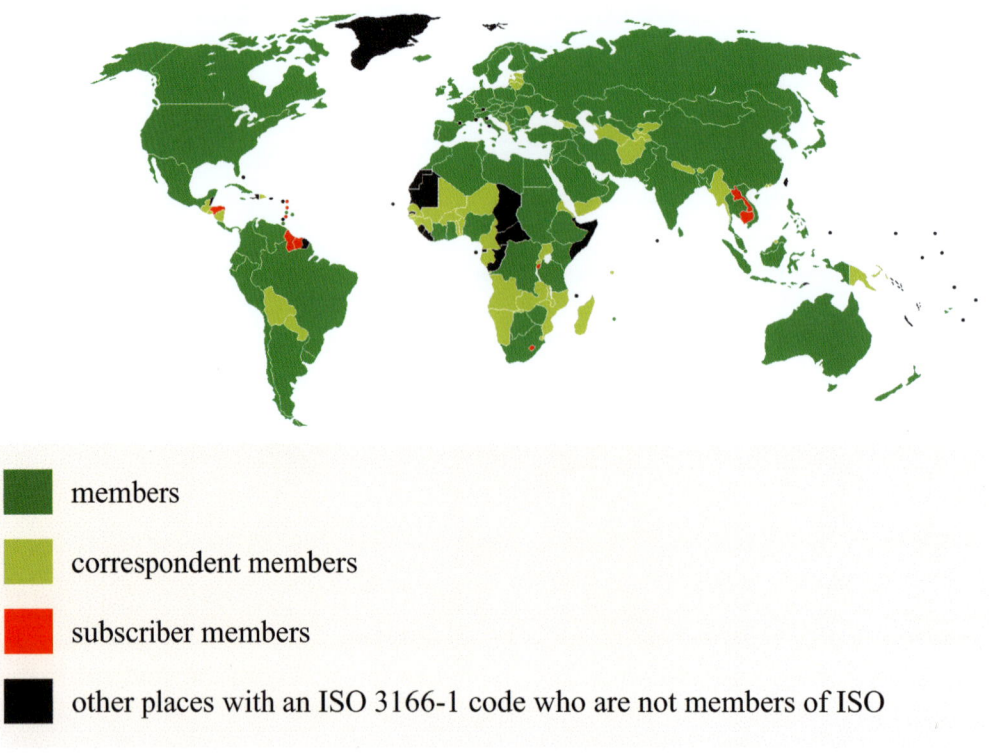

![green]	members
![light green]	correspondent members
![red]	subscriber members
![black]	other places with an ISO 3166-1 code who are not members of ISO

Map showing countries following ISO standards to various degrees

Good Clinical Practice (GCP) is another quality system used in biotechnology. GCP is used to regulate the protection of people who volunteer as subjects in clinical trials. This quality system was developed by the International Conference on Harmonization (ICH), which defines standards that governments should use for clinical trials involving human subjects. The guidelines describe how clinical trials should be conducted. This includes definitions of the roles and responsibilities of those running the clinical trials, including the financial sponsors (usually pharmaceutical or biotechnology companies) and the clinical investigators (doctors). Some key roles include safety, regulatory compliance, and human subjects consent and oversight.

Role of Quality in Biotechnology

Regardless of the type of quality system a manufacturing company follows, all systems require annual audits (review of their procedures for consistency and accuracy of record keeping) to ensure compliance with regulations, performance, and to verify that they are up-to-date. These audits may be conducted by employees of the company, usually as an activity of the **Quality Control** or **Quality Assurance** departments, or by external auditors who work for companies that provide certification for ISO or other systems. In addition, audits by the Food and Drug Administration (FDA) may also occur and are not uncommon in the United States. This process helps the FDA ensure the safety of our drug and food supply.

As an example, a manufacturing protocol, or Standard Operating Procedure (SOP), will generally contain the following pieces of information:

- *reagents (chemicals, etc) and equipment required*

- *protocol (recipe) for making the product*

- *who is making the product and when*

- *confirmation that all of the reagents are used within their expiration dates*

- *how the product will be tested to verify it is of high quality*

- *how the product will be stored and for how long (expiration date).*

A standard operating procedure describes in detail each step that needs to be followed and the criteria for decision making in a specific process.

Summary

In summary, biotechnology laboratories used for research and production in the biotechnology industry are highly regulated and controlled. Many organizations and governmental agencies regulate the manufacture of biotechnology products throughout the world. Companies are always looking for ways to produce their biotechnology products more efficiently and with increased quality standards. Thus, innovative science and engineering are constantly being used to enhance biotechnology research and production.

Concept Reinforcement

1. What is the advantage for a company to have a patent on a discovery they make?

2. Who oversees laboratory practices in the US when they pertain to food or drugs?

3. Who oversees the use of recombinant DNA techniques and provides guidelines for their use?

4. What does GCP stand for and what role does it play in biotechnology?

5. List the critical components of an SOP.

Section 1.5 – Challenges and Opportunities in Global Biotechnology

Section Objectives

- Describe some of the opportunities and challenges in the burgeoning field of global biotechnology

- Describe the role of bioprospecting in global biotechnology

Challenges and Opportunities

Global biotechnology has many challenges and opportunities ahead. Further harmonization of regulations and guidelines between different countries to ease the burden of the development of new biotechnology products is a continuing challenge. In addition, the incorporation of biotechnology in developing countries should be a high priority. Exploring new areas for possible novel organisms or biological components that can be harnessed in biotechnology can aid both the developed countries and developing areas where key organisms may be found. Tailoring biotechnology solutions for specific environments and cultures to match the human and environmental resources and needs in that region will be advantageous for all involved.

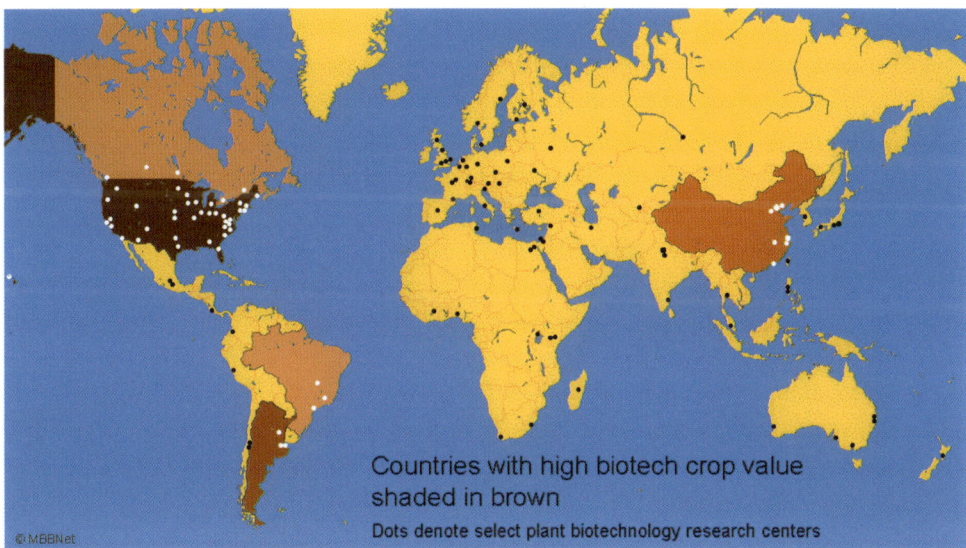

Agricultual biotechnology around the world
Image courtesy of William Hoffman, MBBNet, University of Minnesota

The United Nations Industrial Development Organization (**UNIDO**) helps developing countries and countries with economies in transition to remain or become active in the globalized economy. Its goal is to mobilize knowledge, skills, information, and technology to promote productive employment, a competitive economy, and sound environmental practices in these countries. Biotechnology is an integral part of the UNIDO's plans, as it offers developing countries significant opportunities to address critical social, economic, an industrial problems. Biotechnology can be used to reduce disease and increase food security in developing countries, as well as enhance energy independence.

The key to the spread of biotechnology throughout the world is education. As mentioned earlier, biotechnology is scientifically complex and requires the successful interaction of many different disciplines (**multidisciplinary** approach). Having scientists trained in those many disciplines throughout the world will foster the spread of biotechnology. Consumer education may also foster the spread and acceptance of biotechnology, and specifically genetically modified organisms (GMOs), outside of the United States. The view of the European Union and some other governments, and consumers' perceptions that there are unknown risks associated with GMOs without any obvious benefits to the consumer, has slowed the widespread adoption of genetically modified organisms in the food supply. Further research and education on the effects of GMOs on human, animal, and environmental health with help ease these fears. These may include genetically engineered plants with enhanced nutritional value (i.e. soybeans enriched in omega-3 fatty acids), insect-resistant rice plants, plants engineered for longer storage, crops resistant to drought or other environmental stresses that can be grown in marginally fertile lands, crops engineered to produce better and more nutritionally complete animal feed, farm animals engineered to produce fewer waste products and thus be better for the environment, or crops engineered for particular industrial uses (i.e. improved starch content for making biofuels).

The use of biotechnology solutions for medical and disease problems throughout the world presents a challenge to medical research – finding those targets for disease that can be treated with a drug or therapy, whether it be pharmaceutical or biotechnology in origin. This will rely on continued scientific research into the cause, prevention, identification, and treatment of various diseases and infections.

Bioprospecting

Bioprospecting has occurred for decades, in the sense that drugs used to treat various diseases have arisen from knowledge gained through study of traditional medicines. This knowledge was generally passed down from generation to generation and utilized local plant and animal substances. Crude or semi-purified extracts of plants, animals, microbes, and minerals were often used to treat human and animal diseases. Many antibiotics were first identified in native microorganisms growing in the soil and environment and we continue to find new ones in this way. The *Streptomyces* species has been very useful in the search for antibiotic compounds. Modern bioprospecting takes place over much larger distances and in very remote areas of the globe, by adventurous scientists looking for novel compounds that can be used in medicine or industrial applications. Marine invertebrates are becoming increasingly valuable as sources for novel therapeutic compounds. A new compound for treating neuropathic pain has been purified from a marine cone snail toxin, and other agents to treat cancer are under investigation. The goal of scientists currently working in this area is to gain consent from the native populations before prospecting, collaborate with local scientists and universities, and then share the rewards of such research, both financially and through better education and medical care in the region. The expertise of national and regional scientists is invaluable in identifying and developing new natural products.

Coral reef in Florida
Image courtesy of Jerry Reid, US Fish and Wildlife Service

Summary

There are many challenges and opportunities that exist in global biotechnology today that will continue in the future. The ability to harmonize regulations that cover the biotechnology industry around the world will greatly increase the efficiency at which new products and services can help all citizens and not just those in developed countries. Tailoring biotechnology to best fit the culture and environment of each region is a goal worth pursuing and will require the collaboration of many scientists, business professionals, and political agencies and citizen representatives of diverse cultures and perspectives.. Genetically modified crops and bioprospecting show vast opportunities for global biotechnology, allowing for sustainable food and energy sources around the world, as well as novel biotherapeutics to treat human and animal diseases.

Concept Reinforcement

1. What are a few of the challenges facing modern global biotechnology?

2. What are a few of the opportunities facing modern global biotechnology?

3. What is bioprospecting?

Section 1.6 – The Biotechnology Research Laboratory

Section Objectives

- Describe the unique qualities of a biotechnology research laboratory

- Discuss the importance of quality research programs and techniques in biotechnology

- Describe experimental design and its role in the biotechnology research laboratory

Research Laboratory in Biotechnology

The research performed in a biotechnology research laboratory is not so different from the basic research done across universities and campuses around the world. It may be medical or agricultural, but ultimately it is looking to understand the world around us at the molecular level. This information can then be used to solve problems with biotechnological solutions. The science performed in a biotechnology research laboratory focuses on innovation and practical applications.

One major difference between research in an academic institution and a more industrial setting is the highly controlled nature of biotechnology research. A number of governmental and private institutions regulate the research done and provide rules and guidelines to follow, ensuring that the information generated is high quality, reproducible, and accurate.

The hallmark of Good Laboratory Practices (**GLP**) and any quality research program is proper experimental design. The use of controls to monitor and verify results, repetition, proper statistical and data analysis, and good technique all support proper experimental design. The research scientist should be aware of the **hypothesis** to be tested, what variables will be changing (dependent variables), and which variables will be constant (independent variables). In addition, it is important to use the proper samples, equipment and analysis methods. A good experimental design always has a strong and clear objective and the ability to estimate error and distinguish various effects that may occur in the experiment. A good experimental design relies on the proper planning to make sure the necessary reagents and equipment are available before the experiment is actually performed.

> **Good Laboratory Practices**
>
> GLP refers to a system of controls for laboratory and research organizations to ensure the consistency and reliability of the results; it provides guidelines for how experiments and data should be planned, performed, monitored, recorded, reported, and saved.

> **Hypothesis**
>
> A suggested explanation for an observation or multiple observations; can be tested through experimentation.

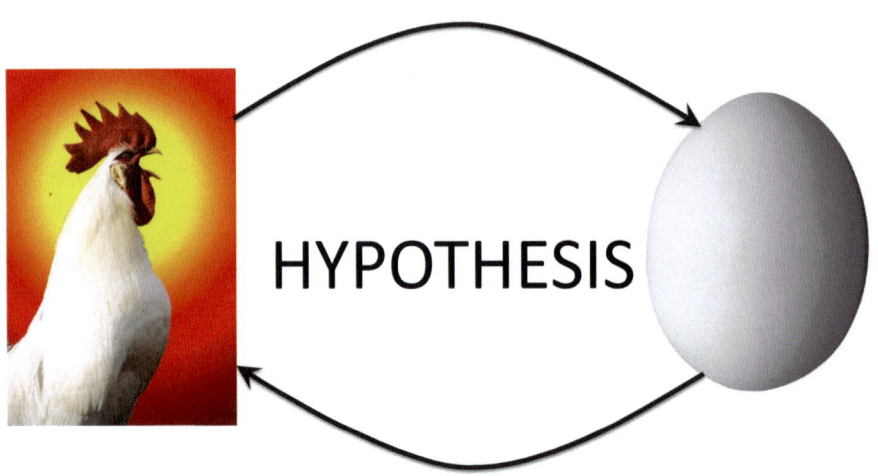

HYPOTHESIS

Developing a scientific hypothesis starts with asking a question and then testing it via experimentation. This then leads to new questions to address.

Biotechnology utilizes many types of tools and equipment to perform research in order to investigate all of the various types of biological molecules in cells and organisms. This investigation relies on the ability of scientists to purify and study the molecules they are interested in. Common tools and techniques for research include nucleic acid purification methods (for DNA and RNA), purification methods for proteins (generally referred to as protein **chromatography**), and subsequent analysis methods. In addition, methods for studying whole cells and organisms may be used, such as microscopy or robotics.

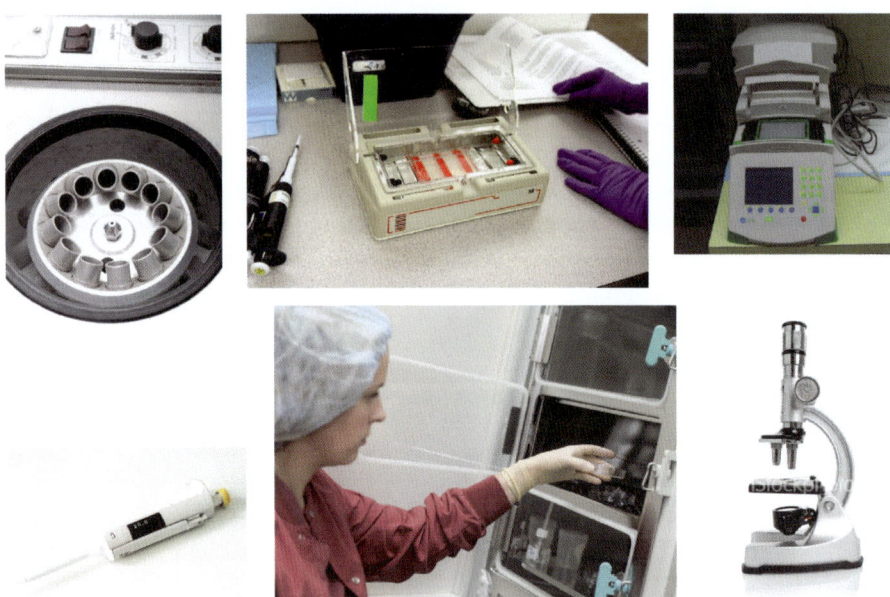

Common equipment found in a biotechnology research laboratory

Summary

Thus, a biotechnology research and development laboratory focuses on innovation and novelty, while maintaining a practical balance with the overall objective of producing a product or process useful to humans. The biotechnology research laboratory is highly regulated and focuses on good laboratory practices (GLP), strong experimental design and proper data analysis. Common tools used in a biotechnology research laboratory include the purification and analysis of various biomolecules such as DNA, RNA, and proteins.

Concept Reinforcement

1. What are the elements of a good experimental design?

2. What does the biotechnology research laboratory focus on?

3. Name a few of the common techniques used in the biotechnology research laboratory.

Section 1.7 – Biotechnology Production Laboratory

Section Objectives

- Compare and contrast a research laboratory with a production (biomanufacturing) industrial laboratory

- Discuss the regulations involved with manufacturing a biological product and the role of contract manufacturing organizations (CMOs)

Production Laboratory in Biotechnology

Biotechnology companies combine business, science, legal, and manufacturing expertise to develop and produce products or services. One of the most important parts of a successful biotechnology company is the capability to manufacture the biotechnology products that it develops through research and through acquiring intellectual property from other scientists. Many discoveries that end up in industrial development are the result of basic research that is performed in university or nonprofit research institutions, or their own research and development (R&D) laboratories. Manufacturing in this environment, unlike some other industries, is highly regulated, monitored, and controlled as mentioned earlier. The manufacturing procedures are highly complex and detailed, requiring the production scientists and engineers involved to focus on quality and documentation. This attention to detail dramatically improves the efficiency and reproducibility of the production. This also assures the exceptional quality of the product, which may be used in the treatment of humans, in which case purity and identity are especially critical.

Although the research environment in biotechnology is highly regulated, primarily through the use of **GLP**, the manufacturing environment is controlled by ever more stringent **GMP** requirements. In addition, the scale of research and scientific techniques utilized in manufacturing and industrial production are significantly larger than that of the discovery phase. For example, a novel enzyme might be identified and studied in a research laboratory using milliliters of the protein, while the transition to the production laboratory may require the protein be purified from hundreds of liters. Thus, the difference in scale is many thousand-fold. This requires vastly different equipment and personnel expertise.

GLP: Good Laboratory Practices

GLP refers to a system of controls for laboratory and research organizations to ensure the consistency and reliability of the results; it provides guidelines for how experiments and data should be planned, performed, monitored, recorded, reported, and saved.

GMP: Good Manufacturing Practices

A series of regulations by the FDA under the authority of the Federal Food, Drug, and Cosmetic Act that require manufacturers, processors, and packagers of drugs, medical devices, some food, and blood products take proactive steps to ensure that their products are safe, pure, and effective.

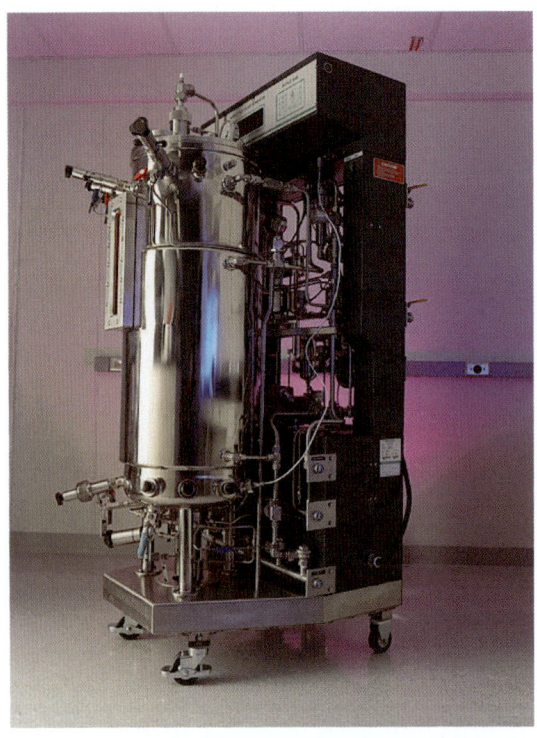

Bioreactor for cellulosic ethanol research. The mobile pilot-plant fermentor shown here has a 90-L capacity and currently is used to generate large volumes of cells and cell products such as outer-membrane vesicles under highly controlled conditions. Future generations of fermentors will be more highly instrumented with sophisticated imaging and other analytical devices to analyze interactions among cells in microbial communities under an array of conditions.Image courtesy of the US Department of Energy

An analytical chromatography system used in a research laboratory

The manufacture of biotechnology products occurs throughout the world and is not restricted to the United States. Regardless of the site of manufacturing, there exists a series of standards that most companies must adhere to in order to supply high quality product for either their own company or for other companies to sell.

Contract Manufacturing Organization

Contract Manufacturing Organizations (CMOs) manufacture products for other companies, particularly in the pharmaceutical and biotechnology industries, because the equipment and expertise needed is vast and very expensive. There are numerous CMOs throughout the world that manufacture of wide variety of biotechnology products.

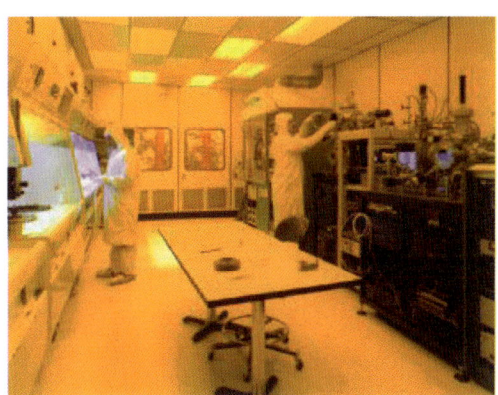

Example of a biomanufacturing process requiring a sterile and controlled environment

> **Contract Manufacturing Organization (CMO)**
>
> A company that supports the development and commercialization of products, primarily for the pharmaceutical and biotechnology industries, by providing manufacturing facilities and services.

Summary

The work done in a biotechnology production laboratory involves the large-scale manufacture of various types of biotechnology products. The environment in a production laboratory is highly regulated, even more so than a research laboratory, and is guided by good manufacturing practice (GMP). Some biotechnology companies have the manufacture of their products done by a Contract Manufacturing Organization, as they may not have the expertise to make the product or the facilities necessary.

Concept Reinforcement

1. What are the two major differences between work done in a research laboratory versus a production (manufacturing) laboratory?

2. What is a contract manufacturing organization?

3. What regulations and guidelines does a production laboratory follow is they are making diagnostic or therapeutic products?

Section 1.8 – Biotherapeutics and Their Role in Biotechnology

Section Objectives

- Describe biotherapeutics and the role they play in biotechnology

- Discuss how biotherapeutics are different from nutraceuticals and probiotics

- Explain the role of bioequivalents or biosimilars in biotechnology around the world

Neutracueticals, Probiotics, and Biotherapeutics

Biotechnology encompasses many different areas and types of products, including foods and food additives (such as herbal supplements, **nutraceuticals** and functional/**probiotic** foods), enzymes and reagents for molecular biology, bioenergy sources, genetically modified organisms (GMOs), and **biotherapeutics**. The area of biotherapeutics has grown dramatically over the past 20 years and will continue to grow in the future due to the increased research being carried out in biotechnology and pharmaceutical companies around the world. In addition, the success of past biotherapeutics has spawned interest in developing novel biotherapeutics for new medical indications. Consumer demand is also driving the industry.

Herbal supplements are a form of dietary supplements regulated by the Food and Drug Administration (FDA). These products must meet specific labeling requirements and must be manufactured in accordance with the FDA's current Good Manufacturing Practices for dietary supplements. Nutraceuticals are not a special category of products in the Food and Drug Act. They are a more general category of product with a coined name that implies a "pharmaceutical" benefit from nutrients or other natural ingredients in foods or dietary supplements. The category does not include drugs or cosmetics. The term "nutraceutical" is generally used in advertising and not on the labels of products.

Under the Dietary Supplement Health and Education Act of 1994 (DSHEA), the dietary supplement manufacturer is responsible for ensuring that a dietary supplement is safe. The Food and Drug Administration is responsible for taking action against any unsafe dietary supplement product after it reaches the market. Manufacturers must make sure that product label information is truthful and not misleading. Ideally, you would choose a product that has been included in a good clinical trial and has proven benefits.

Probiotics include "live" foods or additives that contain live cultures of beneficial bacteria and/or yeast organisms. They are considered healthy for consumption by humans and animals. The most common types of microorganisms used are lactic acid bacteria because they convert sugars, such as lactose, and other carbohydrates into lactic acid. This provides the characteristic taste of fermented dairy foods, in particular, but also acts

---

Nutraceutical

Nutrients and natural ingredients that are deemed to have a therapeutic benefit, similar to pharmaceuticals.

Probiotic

Live microorganisms which, when administered in adequate quantities, confer a health benefit to the host (World Health Organization definition).

Biotherapeutic

Drugs or medical treatments that are based on a biologically derived molecule, usually complex in nature, that are often purified directly from a biological source or expressed in a biological host.

as a preservative in food because it lowers the pH. In addition, probiotic cultures are intended to assist the body's naturally occurring intestinal flora to maintain itself or to help repopulate the GI tract following treatment with antibiotics. Probiotics may strengthen the immune system as well. The most common probiotics are found in dairy products and fortified foods. They may also be found in tablet, capsule, and powder forms. Probiotics, and other food supplements, are regulated by the Food and Drug Administration (FDA) in the US and other countries have similar regulatory agencies.

Many yogurts now contain beneficial microorganisms called "probiotics." ARS (Agricultural Research Service) scientists are developing probiotic bacteria that may lower blood pressure and protect dairy foods from harmful microbes. Photo courtesy of Peggy Greb, US Department of Agriculture

Biotherapeutics encompasses those drugs or medical treatments that utilize a complex biological molecule. This is in contrast to most current drugs, which are small molecules that are chemically synthesized, can be manufactured in large quantities and are easily converted into generic forms when the patent for the original drug expires. Biotherapeutics are most often (though not always) protein in nature and have highly complex structures that are required for biological activity. The complex nature of biotherapeutics is often dictated not only by their biological functions, but also by the complex expression and purification manufacturing schemes they require.

Examples of such biotherapeutics include protein growth factors such as insulin and human growth hormone, protein monoclonal antibodies that are used to treat a variety of medical conditions, vaccines (bacterial, viral, or cancer), somatic or **stem cell** therapies, and **gene therapy**. Antibodies are by far the most common biotherapeutic molecule used to date and consist of specialized proteins of the immune system that can very specifically recognize a particular feature on another biomolecule (usually another protein) and bind to it. The binding of the antibody to its target can have a number of different actions, but often prevents the target from doing what it is designed to do.

Stem Cell

Specialized cells characterized by the ability to renew themselves through mitotic cell division and differentiation into a diverse range of cell types; two broad types of mammalian stem cells are embryonic stem cells that are isolated from the inner cell mass of an embryo or adult stem cells that are found in adult tissues.

Gene Therapy

Insertion of a gene into a person's cells to treat a disease, usually replacing a protein that is either absent or abnormal, through hereditary defects. Gene therapy may use plasmid or viral vectors for gene transfer.

Bioequivalents or Biosimilars

There is interest in those companies that produce generic small molecule drugs to be able to produce generic versions of biotherapeutic drugs, called **biosimilars**, **bioequivalents**, follow-on biologics, or biogenerics. Biotherapeutics, or biological drugs, generally exhibit complex chemical structures and are of high molecular weight. They may be quite sensitive to the conditions and procedures used to manufacture them, since they are either purified directly from a biological source or are expressed in and purified from a host biological source (usually bacteria, yeast, insect cells, or human tissue culture cells). In addition, difficult to detect impurities or breakdown products of the biologic may have serious health implications. Thus, there is concern that copies of biotherapeutics might perform differently than the original patented branded version of the product. In the European Union, a specifically adapted approval procedure has been authorized for certain protein drugs, termed "similar biological medicinal products" which shows that the new product is comparable to the existing product. In the US, the FDA has taken the position that new legislation will be required to address these concerns.

Summary

Biotechnology involves many different types of products, including foods, such as genetically modified foods, food supplements, probiotics, and nutraceuticals. Each of these types of products is regulated differently by the Food and Drug Administration. New on the horizon for biotherapeutics is the role that generic biotechnology drugs will have. These generic versions of biotechnology drugs, which are most often proteins, are called bioequivelents, biosimilars, or biogenerics. They have yet to be approved in the US, but some have been approved in other parts of the world.

Concept Reinforcement

1. What are the differences between probiotics, nutraceuticals, and biotherapeutics?

2. What is meant by the term bioequivalents or biosimilars?

3. What is the most common biotherapeutic molecule?

Biosimilars (bioequivalents)

Generic version of a complex biotherapeutic molecule or drug. This term is used to describe officially approved new versions of innovator biopharmaceutical products, following expiration of any patents.

Omnitrope: The first biosimilar

The European Commission is responsible, along with the European Medicines Agency (EMEA), to design guidelines and approval processes for generic versions of biologic drugs. In January of 2006, they approved the first generic biologic drug (biosimlar), called Omnitrope. This protein is used to treat growth disturbances and growth hormone deficiency in children and adults. Omnitrope was shown in studies to be comparable in quality, safety, and efficacy to the original product called Genotropin. This is the first of what will probably be many biosimilars approved in Europe and possibly the US.

Section 1.9 – Safety Precautions in the Biotechnology Laboratory

Section Objectives

- Describe why safety precautions are important for working in a biotechnology laboratory

- List and discuss several important safety measures, such as personal protective devices, that are used in the biotechnology laboratory

Safety Precautions in the Laboratory

Safety is important in the laboratory setting. The precautions you take to maintain safety are an essential part of your laboratory **protocol**. In industry settings, safety in the laboratory is highly regulated by groups internal and external to the organization, and groups from outside the organization. Most biotechnology companies have their own safety departments that train personnel, monitor work conditions, and dispose of hazardous waste in the proper way. The ultimate goal is your protection. The best way for you to be safe is to be aware of what you are doing and of the safety concerns that exist for each procedure or activity.

> **Protocol**
>
> An established rule or recipe to follow when carrying out a task or an experiment; usually very detailed directions for performing a specific activity in the laboratory.

Before moving on to special equipment and procedures for your safety in the lab, we can start with you at home. Imagine getting dressed to come to work as a research scientist at a biotechnology corporation. What will you choose to wear? What if it is a nice, warm, sunny day? If your company has a casual dress code, you may choose to wear a polo-style shirt, shorts, and a pair of sandals. If you were sitting at your desk, your choices would be just fine. However, if you are going to be doing work in the lab – you should reconsider. You will need something to cover your legs and feet! Long pants or a long skirt and closed-toed shoes are the standard clothing for laboratory work.

Appropriate laboratory attire
(legs and feet covered and safety glasses and laboratory coat on.)

Once you get to the laboratory, some of the basic equipment that scientists use are safety glasses or face shields, laboratory coat, and laboratory gloves. Protecting your hands and face from contact with hazardous chemicals is important, and since you took the time to choose your clothing so carefully before coming to work, you should protect your clothes too. Safety glasses should be able to fit over corrective glasses. For your hands, non-latex gloves are the industry standard, although in many cases latex may be used. Latex gloves offer little protection from chemicals. The pore size of the latex used for gloves is highly variable, and thus chemicals can move right through the glove and make contact with your skin. In addition, many people have or later develop allergies to latex, which in some cases can be quite severe and life-threatening. **Nitrile rubber** or vinyl gloves are common alternatives to latex gloves, and offer better protection, depending on what chemicals you are working with. Each glove manufacturer publishes compatibility charts describing the chemicals against which a particular glove type has been tested so that you can be sure you are using the appropriate glove for the work you are doing.

> ### Nitrile Rubber
> A synthetic polymer used to make laboratory gloves.

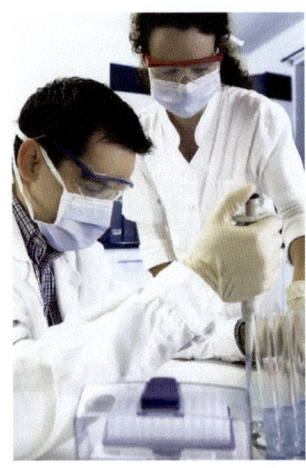

Scientist wearing appropriate safety attire and equipment while doing an experiment: lab coat, safety goggles, and gloves

> ### Microliter
> 1/1,000,000 of a liter; often abbreviated υl.

> ### Reagents
> A reactant or substance consumed during a chemical reaction.

> ### Sublimation
> The process of going directly from a solid phase to a gaseous phase without passing through a liquid phase first.

Much of the lab work in the field of biotechnology is initially done on a small scale (**microliter** amounts of various chemicals or **reagents**) in a research laboratory. Once an appropriate protocol has been worked out, the procedure is often scaled up for manufacture in a manufacturing or production laboratory. Since the scientist working with (among other solutions and chemical mixtures) molecules such as DNA, RNA, and proteins, she or he must be aware of the temperature at which the reagents are kept. In general, when working with RNA and proteins especially, the scientists will work with their reagents on ice, to help stabilize these often fragile molecules. Further, the scientist may occasionally work with dry ice. In fact, in the biotechnology industry, many of the items manufactured and sold to customers require shipment on dry ice to maintain a cold enough temperature so as not to degrade the product. Unlike most frozen solids, dry ice does not melt into a liquid, but rather changes directly into a gas. This phenomenon is called **sublimation**.

The sublimation of dry ice from solid to gaseous form

Dry ice is extremely cold, since it consists of solid carbon dioxide, which is normally a gas at room temperature (-109F). This is cold enough to crack surfaces, so be sure to never place dry ice directly on a counter or other breakable object. Because of its extremely cold temperature, you need to be careful when handling it; use something like a thick oven mitt or tongs to protect your hands. Remember, you can get burned from things that are too cold as well as things that are too hot! Lastly, be aware that dry ice is carbon dioxide and thus gives off carbon dioxide gas as it sublimates. Exposure to dry ice in a closed space can lead to the build-up of carbon dioxide in the blood, or **hypercapnea**. To avoid this condition, be sure to use dry ice in a well-ventilated space.

Safety in the biotechnology laboratory is of key importance and often involves personnel safety equipment such as safety glasses, laboratory coats, and gloves. Proper experimental planning and preparation is the best safety measure anyone can practice in the laboratory. Any question regarding safety in a biotechnology laboratory should be directed to the safety department in the company or institution where you are working.

> ## Hypercapnea
>
> Also spelled hypercapnia. Hypercapnea is excess carbon dioxide in the blood. It leads to disorientation, flushed skin, panic and hyperventilation, among other symptoms. Hypercapnea is potentially fatal.

Summary

Safety is very important in the biotechnology laboratory, just as in any scientific laboratory setting. Various precautions should be taken to ensure personnel safety, including wearing safety glasses, gloves, and laboratory coats at all times. In addition, proper clothing should also be worn. Special precautions are advised when using biological safety hazards, radioactivity, or dry ice.

Concept Reinforcement

1. What is the disadvantage of using latex in laboratory gloves, as compared to nitrile or vinyl?

2. What are the three main personal protective devices used in a biotechnology laboratory?

3. Imagine you are preparing to work on an experiment in the lab. You will need to have the experiment kept cold, so you are planning to use dry ice. What safety precautions should you take?

Section 1.10 – Safety Regulations and Monitoring in the Biotechnology Laboratory

Section Objectives

- List and describe the hazard levels in the laboratory and the handling of hazardous chemicals

- Discuss monitoring of safety regulations by internal or external organizations and why it is important

Biosafety (Biohazard) Levels

There are many regulations and monitoring that occur in a biotechnology laboratory to maintain safe working conditions. **Biosafety Levels** are classification levels for various materials used in the laboratory. Most labs fall under Biosafety Levels 1 and 2. At Biosafety Levels 3 and 4, additional regulations for the laboratory are added to those that have been outlined for levels 1 and 2. Separate regulations and guidelines exist for the use of radioactivity in the laboratory and companies or labs using radioactivity must have a special license to do so and follow strict record keeping and disposal protocols. These radiation safety regulations and guidelines are mandated by the **NRC** (Nuclear Regulatory Commission), as well as OSHA (Occupational Safety and Health Administration) and the EPA (Environmental Protection Agency).

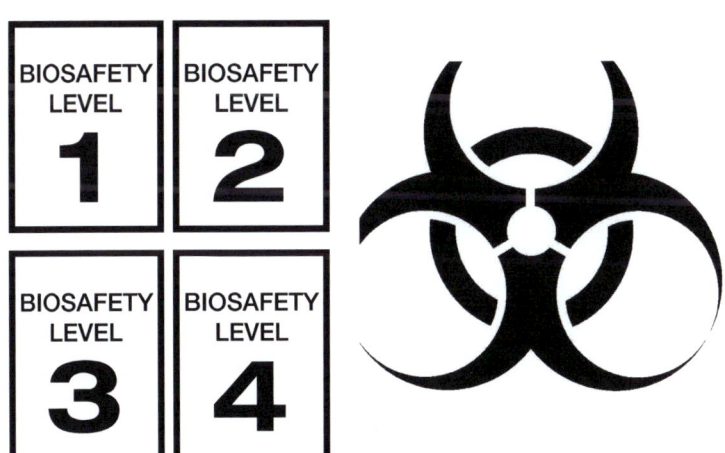

Biosafety level signs and symbol for a biohazard

Biosafety Levels

In the United States, specific biological safety levels are determined by the CDC (Center for Disease Control). They outline the precautions that must be taken when working with various biological agents or biohazards; categorized as levels 1 through 4.

NRC

Nuclear Regulatory Commission.The NRC is part of the US Federal government and is headed by five Commissioners appointed by the President and confirmed by the Senate for five-year terms, with one of them beingdesignated the Chairman and official spokesperson for the NRC; the commission as a body formulates policies, develops regulations governing nuclear reactor and nuclear material safety, issues order to licensees, and adjudicates legal matters.

Safety Precautions in the Laboratory

An example of a procedure that might occur in a biotechnology laboratory is the growth of a pathogenic (disease-causing) strain of bacteria, and subsequent purification of its DNA for sequencing analysis to try to identify the genes involved in the pathogenesis of this particular strain. The growth of such bacteria would require safety equipment, including a **biosafety cabinet** in which the bacteria can be manipulated without fear of being released into the environment or endangering the researcher. In addition, any waste generated needs to be killed by exposure to high heat and pressure in an autoclave. Once the bacteria are destroyed (broken open and killed) to purify the DNA, they are no longer dangerous and the experiment can be carried out on a regular laboratory bench. In addition to a biosafety cabinet and an autoclave, the scientist would always wear a laboratory coat, safety glasses, and gloves.

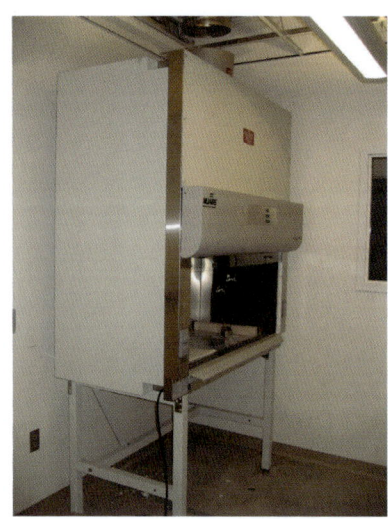

Typical biosafety cabinet used for work with human cells or pathogens
Image courtesy of US Department of Energy

Where would someone look to find out what properties of the chemicals and products they are working with? The biotechnology industry standard is to provide **Material Safety Data Sheets** (MSDS) for each product distributed to a person or laboratory. Many labs maintain a single location within the lab for all MSDS forms so that if someone needed to look up a chemical or product the information is readily available. This is an important safety measure that should not be overlooked. Similarly, as a matter of procedure, you should always review the MSDS before working with a given chemical or other product to verify you are familiar with its uses and hazards.

MSDS symbol

Science Stuff, Inc.

Material Safety Data Sheet

1104 Newport Ave
Austin, TX 78753
Phone/Fax: 1-800-795-7315
www.sciencestuff.com

Nutrient Agar Plates
Catalog #: 1513-10

Hazards Identification

Hazard Description
This product contains no hazardous constituents or the concentration of all chemical constituents are below the regulatory threshold limits described by Occupational Safety Health Administration Hazard Communication Standard 29 CFR 1910.1200 and the European Directive 91/155/EEC, 88/379/EEC, and 67/546/EEC.

Classification System
The classification was made according to the latest editions of international substances lists, and expanded upon from company and literature data.

NFPA ratings (scale 0-4)
Health = 0
Fire = 0
Reactivity = 0

First Aid Measures

General information: No Special measures required.
After inhalation: Seek medical treatment in case of complaints
After skin contact: Immediately wash with water and soap and rinse thoroughly.
After eye contact: Rinse opened eye for several minutes under running water. If symptoms persist, consult a doctor.
After swallowing: If symptoms persist consult doctor.
Information for doctor: Show this sheet.

Toxicological Information

Acute toxicity: none
Primary irritant effect:
on the skin: No irritant effect.
on the eye: No irritant effect.
Sensitization: No sensitizing effects known
Additional toxicological information:
When used and handled according to specifications, the product does not have any harmful effects according to our experience and the information provided to us.

 *Science Stuff, Inc.*

1104 Newport Ave
Austin, TX 78753
Phone/Fax: 1-800-795-7315
www.sciencestuff.com

Handling and Storage

Handling
Information for safe handling: No special measures required.
Information about protection against explosives and fires:
No special measures required.

Storage
Requirements to be met by storerooms and receptacles: <30 C
Information about storing conditions:
Sore in a cool, dry condition in well sealed receptacle.
Class according to regulations on flammable liquids: Void

Exposure Controls and Personal Protection

Components with limit values that require monitoring at the workplace:
This product does not contain any relevant quantities of materials with critical values that have to be monitored at the workplace.

Personal protective equipment
General protective and hygienic measures:
The usual precautionary measures for handling chemicals should be followed.
Breathing equipment: Not required.
Protection of hands: Protective gloves.
Eye protection: Safety glasses.
Body protection: Protective work clothing (lab coat)

This MSDS is for agar plates, which are used to grow bacteria in the lab.

In the case of a chemical spill, the MSDS is crucial, as it will provide guidances on how best to proceed. You should always first check the MSDS, next prepare yourself with the proper personal protective gear, and then clean up the spill. In most laboratories, there is a chemical spill kit available with the necessary tools and products to properly take care of the accident. If you are unsure of what to do, ask someone for help. If the chemical spilled is unknown – you are not sure what it is – you should alert someone else. If, after consulting the MSDS, you determine that the chemical is hazardous, DO NOT attempt to clean it up yourself. Alert someone who is properly trained to handle a hazardous spill and evacuate the area. There should be an evacuation procedure in place. Make sure you are familiar with the procedure so that you can follow it in the event of an emergency. The safety department in most biotechnology companies will coordinate such procedures and train all personnel involved.

The organizations within the biotechnology industry use internal controls to ensure the safety of their employees. **Standard Operating Procedures** (SOPs) are documents that describe how to both prevent and react to emergencies in the laboratory. Most organizations will have a dedicated safety department. It is the job of those working in this department to educate and monitor others within the company to make sure they are following the correct safety procedures. This department is also a resource for emergencies, such as the spill situation described above. Further, the safety group works with external groups such as the **Occupational Safety and Health Administration** (OSHA), Centers for Disease Control (CDC), and the Department of Health and Family Services to ensure that the guidelines set forth by these groups are adhered to for the protection of the worker, both from chemical and biological hazards.

Summary

There are many different types of safety measures performed in a biotechnology laboratory. In addition to personnel safety precautions, laboratories are often classified based on their biosafey level for biological hazards (Levels 1 – 4). OSHA, the EPA, the CDC, and the NRC all regulate safety procedures that must be used in the laboratory to prevent workers from exposure to radioactivity, toxic chemicals, or biological hazards (dangerous bacteria or viruses). The techniques that are performed in a biotechnology laboratory follow standard operating procedures that include details regarding proper safety. Internal safety departments also play a critical role.

Concept Reinforcement

1. How many different biosafety levels are there?

2. What is an MSDS?

3. Imagine that you just walked into your lab. There is a big puddle of clear liquid on the floor and on the lab bench three carboys (large plastic containers for waste chemicals) of some solutions. Two of the carboys are labeled and one is not. Each is about half full. What should you do?

Standard Operating Procedure

A set of instructions that are essentially a directive. When performing a certain task in the biotechnology industry, in order to assure quality, everyone does the task the same way, which is according to the SOP.

OSHA

The Occupational Safety and Health Administration is a part of the United States Department of Labor, whose mission is to prevent injuries, illnesses, and deaths in the work place. It was founded in 1970 as part of the Occupational Safety and Health Act.

Section 1.11 – Professionalism and Proper Scientific Conduct

Section Objectives

- Describe the concept of professionalism and scientific conduct

- Explain the global nature of professionalism

- Discuss the importance of working in teams and respecting differences between people and cultures

Scientific Professionalism and Code of Conduct

The use of scientific information for the good of humans, other living organisms, and the environment depends on the trust that the general public has in the scientific community. Without this trust, the knowledge gained through scientific endeavors cannot be translated into practical solutions to current problems in health care, protection of the environment, species diversity, energy issues, and the general well-being of the planet and its inhabitants. The general public must feel that the scientific community is honest, hard working, unbiased, and has its best interests at heart. Thus, it is the responsibility of every scientist to maintain a high level of standards and professionalism to maintain this trust. Professionalism should overlay all aspects of a scientist's activities, from the design of experiments to the interaction with colleagues, students, legislators, and the general public.

To highlight the importance of this trust, the **National Institutes of Health** (NIH) organized the Committee on Scientific Conduct and Ethics in 1995. This committee consists of members from all branches of NIH and has three main charges:

1. to develop and/or refine guidelines for conduct of research and to develop mechanisms for research ethics training.

2. to develop a course and supporting materials to train principal investigators in leadership, mentoring, and handling conflict.

3. to develop mechanisms to deal fairly and rapidly with allegations of scientific misconduct.

Scientific Conduct and Plagiarism

Scientific conduct involves the actions of scientists and their commitment to honestly and openly share their findings with other scientists for review and discussion, to accurately present experimental design and results, and to use their own words to describe their work. Copying words, ideas, or results from someone else is called **plagiarism** and is a serious offense.

National Institutes of Health

The National Institutes of Health are part of the US Department of Health and Human Services (HSS) which is the primary Federal agency for conducting and funding medical research. The NIH is composed of 27 institutes and centers.

Scientific Conduct

The code of scholarly conduct and ethical behavior in scientific research, whether in academic institutions or business environments.

Plagiarism

The verbatim use of the words or phrases of others as if they are your own. Plagiarism includes claiming or implying original authorship for work that is not your own and is considered scientific misconduct in the field of science.

The critical analysis of scientific findings is what drives the advancement of science through open communication and sharing of ideas. This scientific discourse allows scientists to think critically about their findings and refine their hypotheses to fit what is known to date. Science cannot be done in a vacuum or in the absence of others' findings. It is a collaborative field, where the research of each individual feeds into a pool of knowledge that is constantly being updated, verified, repeated, and refined. Although mistakes can happen, they must be acknowledged quickly and openly to minimize any negative impacts on the relevant field of research. Under no circumstances should results be fabricated or embellished. This not only acts to undermine the credibility of the scientist(s) involved, but also the scientific community as a whole.

Biotechnology is no different than any other field of science. It requires those involved to behave in a professional manner with the utmost of integrity and collaboration, while sharing their findings with scientific colleagues and the general public.

Iwan Petrowicz Pawlow

Robert Koch

Alexander Fleming
Image courtesy of US Government

Francis Collins
Image courtesy of US Government

Robert Ballard
Image courtesy of NOAA

Darleane Hoffman
Image courtesy of Lawrence Berkeley National Lab

Pat Cowings
Image courtesy of NASA

Famous scientists throughout history, all of whom exhibited proper scientific conduct and professionalism

Summary

Proper scientific conduct is key to maintaining the public trust in scientists and their findings. This involves honest, hard working, and unbiased collaboration, including sharing results with the public following critical peer review. Findings must be novel and unmodified from the original analysis, explanation of the result and conclusions, as stated by the scientist who performed the experiment. The role of a scientist is to be professional at all times.

Concept Reinforcement

1. What is the link between scientific conduct and the general public?

2. What is plagiarism and how does it apply to scientific conduct and scientific professionalism?

3. What other types of problems would be considered scientific misconduct?

Section 1.12 – Experimental Design and Data Analysis

Section Objectives

- Describe experimental design and proper data analysis

- Discuss commonly used statistical terms in data analysis

- Explain the interdisciplinary nature of global biotechnology and its impact on the scientific process

Experimental Design and Data Analysis

The heart of scientific research is proper experimental design, or how experiments are planned, conducted, and analyzed. As indicated, it is a total plan for all aspects of an experiment, from the question to be addressed, the parameters to be measured, the samples to be tested, the number of replicates to be processed, and how the results will be analyzed. A thorough experimental design will maximize the likelihood that interpretable results will be obtained. Often the analysis will involve the use of proper statistical methods that are appropriate for the data set generated. In general, at least three (3) measurements for each parameter must be taken so that the variation around that measurement can be assessed (**standard deviation** or **coefficient of variation**). This gives scientists information as to how **precise** their measurement technique is. Ideally, you would want measurements to be precise, as this allows for stronger confidence in the results. This is in contrast to how **accurate** a measurement is, as this implies how close the experimental measurement is to an established standard. Often standards are not available for all techniques or measurements made in science or biotechnology. The **National Institute of Standards and Technology** (NIST) provides many standards that can be used in biotechnology.

Designing an Experiment

Let us design an experiment to investigate the effect of sunlight on plant growth. Our hypothesis is that sunlight is necessary for green plants to grow. How would we test this hypothesis? What would we need to perform the experiment? What controls should we use? One possible way to test this hypothesis is to transplant multiple kinds of plants into small pots and then cover half of the plants with boxes that block the light, and leave the other half of the plants uncovered in the sun. Each plant type would need to be tested under both conditions in multiple numbers (i.e. replicates of 3-5). This will allow us to detect an average response and standard deviation, since individual plants may vary somewhat on their response to light or no light. To monitor the effect of no light on plant growth, we could measure each plant's height at the same time every day and then plot these measurements on a graph as plant height (in centimeters) versus time (in days). In addition to providing the plants with the same dirt and a container, they will also need

Standard deviation (SD)

The most common measure of statistical dispersion, which is how close replicate measurements from a single sample are to each other. SD is calculated by taking the square root of the differences between the individual values and the mean squared, divided by the number of replicates.

Coefficient of variation (CV)

The ratio of the standard deviation to the mean (average; i.e. divide standard deviation by the mean). CV is useful for looking at the variation of replicate measurements, which are often expressed as percent CV.

Accuracy

How close a measurement falls to a given established standard, often obtained from the National Institute for Standards and Technology (NIST).

Precision

How close a series of individual measurements are to each other.

to receive the same amount of water every day, as well as identical amount of the same plant food. This would result in the exposure to light as the only variable parameter that is different between the two sets of plants.

Plant grown with exposure to sunlight

At its best, science involves the interplay of scientists from around the world. Given the advances in communication technology, this is much easier than it used to be. Modern scientists interact not only with those in their departments, but other departments in the university or company where they work, and with scientists from other universities or companies across the globe. This **interdisciplinary** method of research not only increases the speed of scientific discoveries, but also the rigor applied to their review. Critical analysis is the key to good scientific method. The interdisciplinary nature of science requires that researchers are sensitive to the different national, cultural, language, regulatory, and time factors associated with different parts of the world and respectful of them.

Particularly in the field of biotechnology, the ability to work effectively in teams is not only critical to success of the team project, but also to your career in the field. A single biotechnology product requires the involvement of basic research and development (R&D) scientists, production or manufacturing scientists and engineers, quality control and quality assurance representatives, marketing experts, a sales force, packaging and shipping experts, and a project manager to oversee the entire effort. This is in addition to upper management,who provide the team members with the guidance and tools they need, to be successful. Others involved in the team might be legal representatives to determine whether the technology to be developed has already been patented or bioinformatics specialists if the product involves complex data analysis.

An international scientific conference.

Summary

Proper experimental design and data analysis are key to any scientific endeavor. Planning for an experiment involves defining the question or problem to be addressed, what reagents and equipment are required to perform the experiment, and how the results will be analyzed. Once completed, the results should be shared with other scientists for critical review. Various statistical parameters can be used to analyze the results, including average, standard devation, accuracy, and precision. Most scientists work in teams utilizing an interdisciplinary approach.

Concept Reinforcement

1. What is experimental design?

2. What is the difference between precision and accuracy?

3. What role do teams play in biotechnology and who might sit on a product development team?

Science, Statistics, and Data Analysis

Statistics is the science of collecting, organizing, analyzing, and interpreting numerical data or experimental results. It is valuable, not only in the biological sciences, but also in economics, education, and social policy as well. Training in statistics is a key element in any scientist's education because it allows them to learn to define problems, think critically, design experiments, and draw conclusions from the results. Statistics also allow scientists to evaluate uncertain results with a defined degree of confidence or probability.

Section 1.13 – Effective Oral and Written Communication

Section Objectives

- Define effective communication practices, both for written and oral communication

- Describe what a storyboard is and how it is used in communication

- Discuss the advantages and disadvantages of using PowerPoint as a communication tool

Effective Communication

The ability to communicate with individuals with diverse backgrounds effectively can determine the success or failure of a team project or new biotechnology product. In order for a new idea to become a practical application, scientific and business ideas must be shared between people with very different backgrounds and levels of expertise, such as scientists, lawyers, business professionals, legislators, or funding agencies (such as those providing grant money like the government or those providing private funds for research such as banks or venture capital firms). Without the ability to effectively communicate complex scientific information, a new technology, product, or company may fail.

Effective communication may be in written or oral format. Both are critical to different aspects of team projects or communication in general. Effective communication is at its simplest the use of written or oral language or graphics to concisely and accurately convey information or ideas to others. There are a number of guidelines that can increase the likelihood that any type of communication will be effective and inform, persuade, and/or entertain the audience to which it is directed.

> **Effective Communication**
>
> The ability to convey ideas or information in a concise and accurate way to others such that the purpose of the communication is achieved.

Chelsea Clinton speaking to a public audience

The first step to effective communication is to understand the purpose of the communication. Who is the intended audience? Is the presentation or paper supposed to entertain, inform, or persuade the audience? Does the audience consist of peers (scientists presenting to other scientists), non-peers, others in a company, customers, media representatives, legislators, or is it a broad mix of the general public? Understanding the purpose of the communication and the audience can define the style in which the information will be shared, as well as the focus and scope of the communication. What types of presentation tools are available (overhead projector, slides, computer with a projector and Power-Point, printed copies of text or figures, or text and/or figures online)?

In general, communicate at least initially to the least advanced members in the audience on the particular topic. Quickly get them up to speed on the topic before jumping into the details. This will allow a quick review for those already versed in the field and new information for those not familiar with the field.

Effective communication should fit the circumstance and the audience. In general, for both oral and written communication, a title, introduction to the topic, body of the information, conclusions, summary, and references should be included. This is particularly true for communications that are meant to inform or persuade. References are critical to effective communication, as they can point audience members to other sources of information if they would like to know more. It also reaffirms that no plagiarism is occurring. One good rule of thumb for effective communication: tell the audience what you are going to tell them, tell them the information, and then remind them of what you told them. It may seem redundant from the presenter's perspective, but turns out to be highly effective from the audience's position. These points might be considered in an abstract or executive summary, in which the purpose of the communication is to summarize information and conclusions briefly, but thoroughly.

The use of illustrations such as pictures, figures, or types of graphical data can be quite effective in conveying information and highlighting points that you want to make. Make sure that they are relevant to the topic and integrate with the flow of the paper or talk. The most difficult aspect of any type of communication is the flow between ideas. Be sure to give extra time and thought to the transition phrases used in the communication, whether it be oral or written.

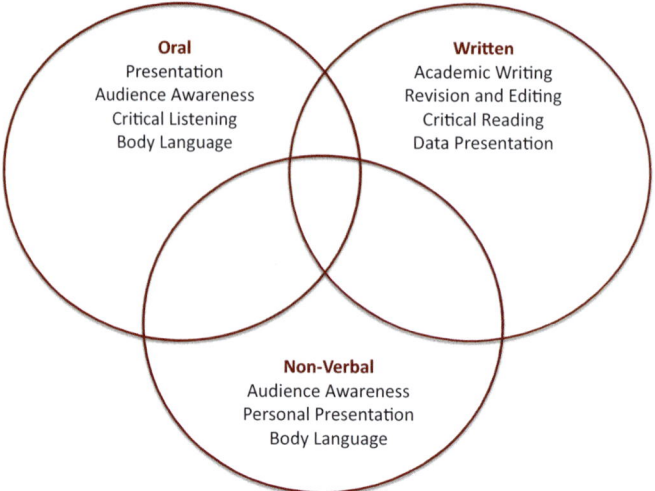

Important Communication Skills

Oral
Presentation
Audience Awareness
Critical Listening
Body Language

Written
Academic Writing
Revision and Editing
Critical Reading
Data Presentation

Non-Verbal
Audience Awareness
Personal Presentation
Body Language

Storyboarding

To be most effective, the overall communication should tell a story. Often the first step in creating an effective oral or written communication is to **storyboard** the ideas that you would like to cover and the illustrations that will go with them, as well as any key transition phrases you would like to use. This can be the most difficult part of the creative process, but it also allows for the most latitude in ideas at the beginning, when the presentation is not completely defined.

Example of a storyboard format that is useful when designing, planning, and generating an oral presentation

Microsoft PowerPoint and Oral Presentations

The use of Microsoft PowerPoint to generate and present oral communications has become more and more commonplace in both business and academic settings. PowerPoint can be a very powerful medium for effective presentation. However, if not used well, it can be a major distraction that interferes with successful communication. Information on each slide should be limited to one or two main points, illustrated using minimal text and few graphics. Verify that the font size and colors used can be seen from the back of the presentation room with the lighting available. Do not use movement in the presentation (moving text or graphics with sound effects) unless absolutely necessary, as they can be quite distracting if not used properly. Preview slides in the presentation room on the equipment provided before they are projected, because PC and Mac computers will likely show text and symbols differently when projected in PowerPoint. Make sure any inserted functions (websites, movies, video, sound, etc.) work properly before giving the presentation. If the PowerPoint presentation is to be used as an information resource for the audience at a later time, then more text may be included in the slides or in the notes section. In this case, it is critical that key references are included in the presentation.

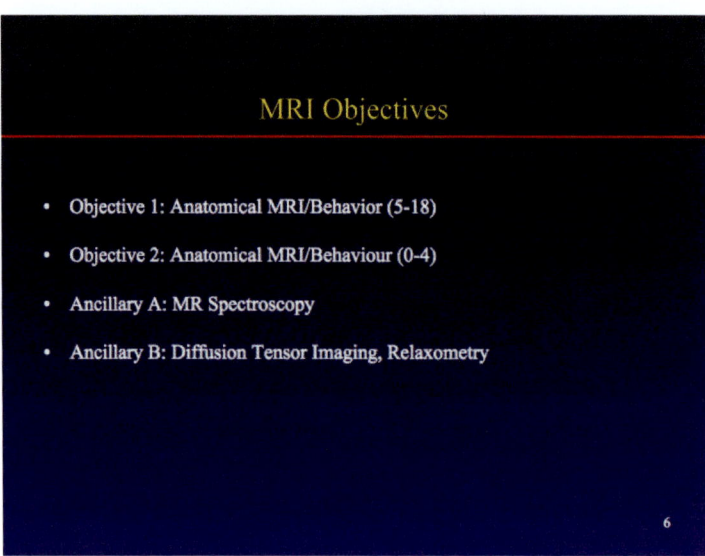

MRI Objectives

- Objective 1: Anatomical MRI/Behavior (5-18)
- Objective 2: Anatomical MRI/Behaviour (0-4)
- Ancillary A: MR Spectroscopy
- Ancillary B: Diffusion Tensor Imaging, Relaxometry

6

Target Identification
Selection Criteria

Metrics	Selection Criteria	Ideal Firm to Acquire
Business Profile	Business Philosophy	Values customer service and agents
	Business Model	Traditional Brokerage Model
External	Location of Offices	Covers majority of Greater Cincinnati Area
	Market Segment	Targets high end market segment
	Market Share	Greater than 4%
Internal	Sales volume per Agent	High grossing agents
	Feasibility of Acquisition	Open to acquisition

Examples of good and bad PowerPoint slides in an oral presentation

Effective communication is involved in all aspects of biotechnology, whether it be at the research bench to the legal department in preparing a patent application or in the marketing literature that is given to customers touting the advantages of a particular product or treatment. Science can only progress and maintain the public trust if the critical interpretation of the information is conveyed to other scientists and the general public in a meaningful way.

Summary

Effective communication, whether it be oral or written, starts with a good plan. What is the purpose of the presentation, who is the audience, what types of information need to be presented, and what is the best medium to use? All of these issues should be addressed before the communication is presented to enhance its effectiveness. The use of a storyboard can greatly enhance the overall effectiveness of a presentation or written document since it allows for the material to be organized in a logical manner and encourage the presenter to focus on clarity.

Concept Reinforcement

1. What are some of the keys to effective communication? Which apply most to the types of communication you currently do in school or other groups?

2. What is a storyboard and how are they used in effective communication?

3. What are some advantages and disadvantages of using PowerPoint for communication?

Section 1.14 – Scientific Careers in Biotechnology

Section Objective

- Describe the variety of career paths in science that exist in biotechnology

Education Requirements in Biotechnology

The field of biotechnology has exploded over the past three to four decades, since the birth of modern molecular biology in the 1970s. Thus, the career opportunities in biotechnology are also very promising. Although most jobs in biotechnology involve a strong science background (biology, molecular biology, biochemistry, chemistry, agronomy, engineering, mathematics, physics, or computer science), not all require proficiency in science. It is important to have some scientific experience when practicing in the field, and their experience also increases the likelihood of advancement. This section will focus on the scientific careers in biotechnology.

The most obvious careers in biotechnology are those working directly at the laboratory bench, that is, designing experiments, performing experiments, and analyzing results for basic research and development (R&D). These scientists are the innovators who launched biotechnology and who continue to drive the field forward. An extremely strong background in a scientific field is a requirement, and often this involves either a **Master's of Science** (MS) degree or a **Doctor of Philosophy** (PhD) degree. These degrees are pursued after earning a bachelor's degree and are offered by universities around the world. The typical R&D scientist will research new areas for potential technologies, products or medical applications as dictated by the focus of his or her company. Basic research in biotechnology conducted in a university setting allows more latitude in the areas investigated. This work generally requires approval and funding from the federal or state government **grants**. R&D scientists are often dedicated and intelligent individuals with a strong drive for discovering new information, solving problems, or finding new uses for old ideas. Scientists with a **Bachelor of Science** (BS) or an **Associate's** degree in a scientific field might find themselves working at a starting level position in an R&D laboratory, so an advanced degree is not absolutely necessary, but it is highly desirable, especially when opportunities for professional advancement are concerned.

> **Grants**
>
> Financial funding obtained through a competitive process with the federal or state government or charitable foundation; potential research projects are evaluated and only the most promising and well-planned are generally funded.

Degree	Description
Associate's Degree	A two-year degree usually awarded by a technical college. The focus is on the technical aspects of a career, such as a lab technician.
Bachelor of Science (BS)	An undergraduate academic degree awarded by colleges and universities upon completion of a specific set of coursework requirements.
Master's of Science (MS)	An advanced academic degree that usually applies to the sciences and sometimes the social sciences; involves advanced coursework and completion of an original research project, but not as involved or complex as the PhD project; the time frame for completion is usually less than for a PhD degree.
Doctor of Philosophy (PhD)	An advanced academic degree that applies to graduates in a wide variety of disciplines, in both scientific fields and the humanities. A candidate must prepare and submit a thesis based on their own original research and then defend this work in a formal setting. A PhD generally also requires the completion of advanced coursework over an extended time frame (4-7 years of post-graduate work).

Scientific Career Options in Biotechnology

R&D scientist in a biotechnology laboratory

SOP

Standard Operating Procedure. A set of instructions that are essentially a directive. SOPs are followed when performing a certain task in the biotechnology industry. In order to assure quality, everyone does the task the same way according to the SOP.

In addition to bench laboratory positions in R&D, those with expertise in bioinformatics and computer programming may also work in the product development phase of biotechnology. They may also be involved in many aspects of developing new technologies and products, from data analysis and management to writing programs for customer use. This is also true of engineers, as they may be involved at multiple steps in the process of developing a biotechnology product.

A similar laboratory bench position to an R&D scientist is a production or manufacturing scientist. Like the R&D scientist, the manufacturing scientist also designs, performs, and analyzes the results of experiments, but generally with less freedom to stray from standardized and optimized manufacturing protocols (**SOPs**). Production scientists are the heart of the biotechnology company, because without them, there will be no products available for the customer to purchase. They are highly skilled and dedicated scientists who see the value in following **validated** and previously tested methods for manufacturing a specific product to very detailed **requirements** and **specifications**. They usually have at least a Bachelor's degree, but often a MS or PhD degree in a scientific field like biology, chemistry, or engineering. The field of engineering plays a huge role in manufacturing, particularly in biomanufacturing, since the large-scale equipment and machinery involved are often novel and designed in the company where they are used.

Laboratory column testing of media at the MSE Testing Facility in Butte, Montana
Image courtesy of the EPA

In addition to bench laboratory scientists, people with science backgrounds and laboratory experience are often utilized in scientific technical support and training roles. These individuals provide technical help for customers and may offer training to customers on how to best use the products they purchase Minimally, most biotechnology companies need to provide training for new employees and current employees on new products. Therefore, scientific training departments can be critical in developing and maintaining a highly skilled and efficient work force. Because the field of biotechnology is so technologically advanced and fraught with rapid change, any employee must be kept up-to-date on new techniques and innovations in the field. Those who perform such technical support or training functions most often have at least a BS or MS degree, but may also have a PhD degree. In addition to training customers or internal scientific staff, biotechnology companies or departments within a university may have outreach programs whose mission is to provide biotechnology-related educational opportunities to younger students and their teachers, as well as the general public. These outreach programs are often staffed by trained scientists with BS, MS, or PhD degrees and provide a great service to the community. This is another way that scientists can foster trust in the public by willingly sharing scientific information in an open and supportive learning environment.

Outreach education program in biotechnology and science for students

Requirement

A description of what a product must be able to do to fill a customer's needs.

Specification

A detailed description of the physical characteristics or functions of a product; very defined results will be termed acceptable or unacceptable and will be unique to each product or product type.

Validation

Verification that something consistently satisfies a certain set of criteria. Validation is often included in a quality system for manufacturing.

Quality Assurance (QA)

Provides assessment of the quality control conducted to ensure product quality; the activity of providing the evidence needed to establish confidence that quality-related activities are being performed effectively.

Quality Control (QC)

The operational techniques and activities used to fulfill and verify the requirements for product quality; inspection, analysis, and action required to maintain product specifications.

Regulatory Affairs

This group has the responsibility for ensuring that companies follow all of the laws, regulations, and guidelines for their industry by working closely with local, state, and federal governmental agencies.

Scientists are also involved in the **Quality Control** (QC), **Quality Assurance** (QA), and **Regulatory Affairs** departments in biotechnology companies. These scientists monitor and verify that all products manufactured meet the needed specifications and requirements and that all necessary federal and state (as well as world wide if the product is being sold outside of the United States) regulations have been followed and met. They are responsible for the quality and consistency of biotechnology products and they often deal with the appropriate governmental agencies, such as the FDA, USDA, EPA, and others. Attention to detail and strong personal conviction are good attributes for a QC, QA, or regulatory affairs scientist.

Advanced degrees in Biotechnology.

Many colleges, universities, and technical colleges offer advanced degrees in biotechnology or a closely-related field. Many of these programs offer a Master of Science (MS) degree in biotechnology, and some offer a Doctor of Philosophy (PhD) degree as well. These advanced degree programs focus on the science behind biotechnology, but often also relevant business, legal, and ethical issues. Business topics include the ability to market a product and act as a leader in a biotechnology company. Intellectual property and inventions that drive the science of biotechnology are key to the discussion of legal issues. Lastly, ethical concerns in biotechnology often deal with controversies surrounding stem cells, human clinical trials, possible effects of the technologies on the environment, and governmental policies. A diverse and well-rounded approach to biotechnology will enable graduates to quickly and effectively integrate into the biotechnology industry and intelligently converse with the general public, as well as with other scientists, business managers, lawyers, and politicians.

Summary

Thus, there are many types of careers in biotechnology that require a strong background in one or more fields of science with minimally a Bachelor's degree, but ideally a Master's or PhD degree. They include research and development, manufacturing, quality control, or quality assurance scientists. In addition, training specifically in biotechnology, and not just a certain scientific field, can also benefit an employee in any of these positions in a biotechnology company. Scientists are also involved in technical support and training.

Concept Reinforcement

1. What roles do R&D and manufacturing scientists play in biotechnology? What training do they need?

2. What types of scientific positions are available in biotechnology in the field of quality?

3. Are advanced degrees available specifically in biotechnology?

Section 1.15 – Non-Science Careers in Biotechnology

Section Objective

- Describe the variety of career paths in the field of biotechnology that are not science related

Non-science Career Options in Biotechnology

Not all careers in biotechnology require a scientific background. Though, as discussed, many scientists populate the marketing and sales departments of biotechnology companies, not all employees involved in these areas have science backgrounds. They may have more general business experience and then learn to apply it to the field of biotechnology. Unlike general consumer goods (cars, food, electronics, books), biotechnology products are highly technical and generally specialized products. Thus, they require highly specialized market research and marketing campaigns directed to very specific customers. In addition, a biotechnology sales force must also be highly trained on details of the products and their advantages and disadvantages, so that they can educate customers on their proper use. Education and experience in both science and business significantly contribute to excellent credentials for working in the sales and marketing areas of biotechnology.

Along with sales and marketing, most biotechnology companies have people who specialize in business development and/or technology acquisition. These individuals tend to have legal backgrounds, often involving science training as well. They act as facilitators to bring in new technologies from other companies or from universities. Most biotechnology companies have at least one full-time lawyer on staff who specializes in patent law and litigation, but many have full legal departments that focus on the issues related to the technologies that the company develops. These departments may include **patent agents**, who are not necessarily lawyers, but have passed special training requirements (registration examination) and have become licensed to prepare, file, and prosecute patent applications with the **United States Patent and Trademark Office** (USPTO). Unlike attorneys, a patent agent cannot give legal advice or represent a company in front of the USPTO or judicial system.

> ### Patent Agent
> A person who passed the registration examination from the USPTO and is licensed to prepare, file, and prosecute patents with the USPTO.

> ### United States Patent and Trademark Office (USPTO)
> The USPTO is a federal agency in the Department of Commerce. Its mission is to promote the progress of science and the useful arts by securing for limited times to inventors the exclusive rights to their respective discoveries; the office employees over 7000 full time staff to support its major functions – examination and issuance of patents and trademark applications and disseminating patent and trademark information.

79

Other non-scientific career opportunities available in many different types of industries are also found in biotechnology. These include customer service, product packaging and shipping, financial tracking and accounting, information systems support, and others. Strong writing and communication skills enhance credentials for working on grant applications, manuscript preparation, and Standard Operating Procedures.

While not directly involved in a biotechnology company, there are many supporting roles in the biotechnology industry that involve educating, advocating, or lobbying on behalf of the biotechnology community. Science reporters for local, regional, and national newspapers, magazines, radio, television, and internet stations report on the biotechnology industry and help educate the public on what biotechnology means and how it impacts our lives. Those who work for biotechnology trade organizations, such as **BIO** (Biotechnology Industry Organization), act as advocates and champions for the biotechnology industry in general, lobbying the government for favorable legislation.

Summary

The career opportunities in the biotechnology field are diverse and expanding, particularly as biotechnology plays an ever-increasing role in health care and drug development, agriculture, bioenergy, and environmental cleanup. Minimally, a college education is required for most, but not all, jobs in this field and often post-graduate study is preferred with a strong emphasis on science or the field you will be working in directly (law, sales, marketing, finance, etc.). Careers in biotechnology afford the opportunity to be innovative, both in scientific and nonscientific arenas.

Concept Reinforcement

1. What non-science careers are available in the field of biotechnology?

2. What is a patent agent?

3. How do biotechnology companies help in educating the public?

> **BIO**
>
> The Biotechnology Industry Organization is the world's largest biotechnology organization, providing advocacy, business development, and communication services for its members worldwide. BIO's mission is to be a champion of biotechnology and represent its interests.

Unit Two

Section 2.1 – Understanding and Using Chemical Solutions

Section Objectives

- Describe the periodic table of elements and how it can be used to calculate the molecular weight of a compound

- Describe how elements can be combined to create compounds and chemical solutions

- Define buffer and discuss how buffers are used in biotechnology

The Periodic Table

The basis for all solutions that are useful in the field of biotechnology is the periodic table of the elements. The periodic table is a systematic display of all of the chemicals known in their most basic or elemental form, hence the title. The most current (January 2008) table contains 118 elements, most of which are found in nature and a few of which have been synthesized, or created, by scientists.

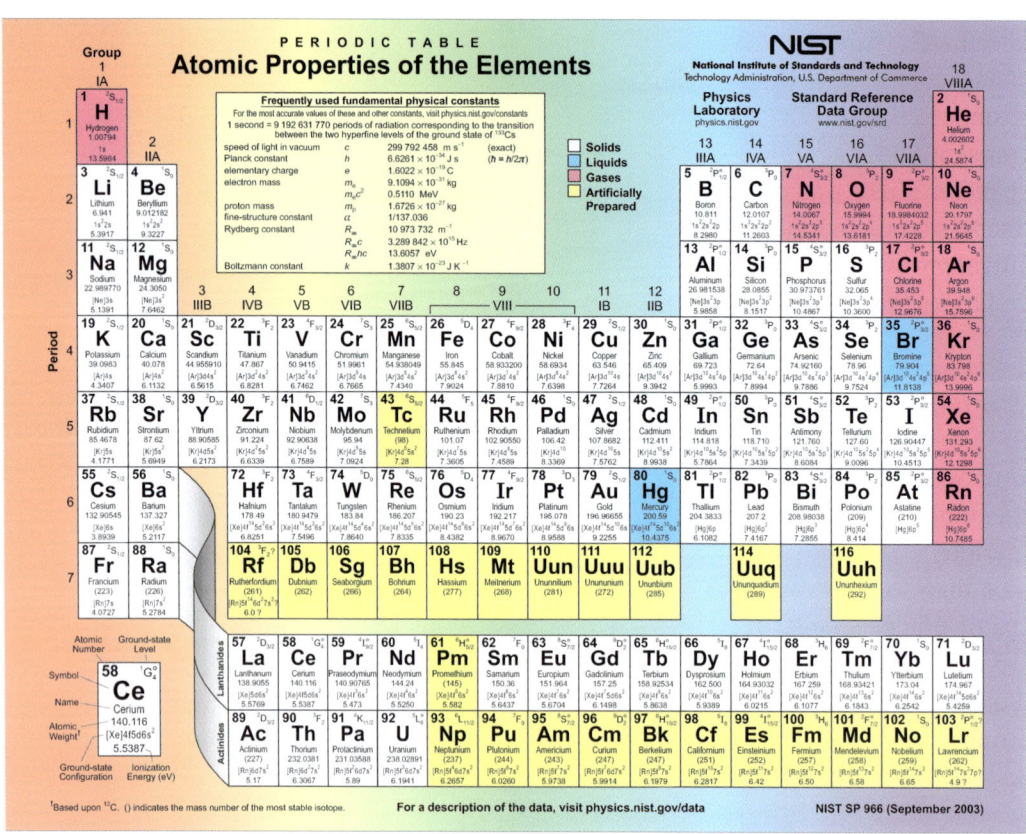

The Periodic Table of Elements (118 total)

Atomic mass
The total mass of protons, neutrons, and electrons in a single atom; generally interchangeable with atomic weight.

Atomic mass unit (amu or u)
Atomic mass units are defined as 1/12 the weight of the carbon-12 isotope, and carbon-12 is defined as weighing exactly 12 amu.

Compound molecule
A substance that has more than one element in a single molecule.

Molecular weight
The weight in grams of a mole of an element or compound; always expressed in grams per mole; weight in grams of 6.022×10^{23} atoms/molecules of an element or compound.

Glucose
A simple sugar with the chemical formula $C_6H_{12}O_6$.

The periodic table is set up in a way that allows one to gain understanding of the properties of each chemical and how it is expected to react when combined with other elements in compound form. A basic understanding of the periodic table is necessary in order to move forward with understanding and using chemical solutions.

The periodic table contains information that is essential to making chemical solutions. Beyond knowing which elements make up a compound, one must also know the **atomic mass** (number of protons and neutrons in the nucleus) of a given element in order to calculate its **molecular weight**. The molecular weight is then used in subsequent calculations to make solutions. For instance, NaCl is a **compound molecule** of table salt. The molecule is made of one atom of sodium and one atom of chlorine. Chemically, table salt is known as sodium chloride. Its molecular weight is the combined atomic mass of one atom of sodium and one atom of chlorine. That is, about 23 + 35.5 which equals 58.5 amu or **atomic mass units (see Periodic Table for the atomic mass of sodium and chlorine)**. Thus, the molecular weight of a substance is the weight in atomic mass units of all the atoms in a given formula. What about more complex molecules, like **glucose**? Glucose is written chemically as $C_6H_{12}O_6$.

Calculating the molecular weight of glucose involves multiplying the atomic mass of each element by the number of molecules present and then adding the values together. Glucose, then, has a molecular weight of approximately 180 amu (6 x 12 for carbon, 12 x 1 for hydrogen, and 6 x 16 for oxygen). The atomic mass of carbon, hydrogen, and oxygen can be found in the Periodic Table.

Note that rounding numbers is allowed when calculating molecular weights. The molecular weight of a substance is needed to tell us how many grams are in one **mole** of that substance. The mole is the standard in chemistry for describing how much of a substance is present. One mole is based on **Avogadro's Number**, which is 6.022×10^{23} molecules per mole. This means that one mole of a substance is equal to 6.022×10^{23} molecules of either a given compound or a given element. Note that molar mass (grams/mole) is the equivalent of atomic mass (amu or υ), but the units are denoted differently – i.e. hydrogen has a molar mass of one gram/mole and a molecular weight of one amu.

Compounds and Solutions

Compounds are chemicals that have more than one element in a single molecule. Glucose, table salt, and water are all compounds. Compounds are different from any of their individual elemental parts, physically and chemically. That is, a chemical reaction takes place that bonds one element to another, creating a new molecule. For instance, in the case of glucose, sugar has completely different properties than oxygen, hydrogen, and carbon. Table salt (NaCl) is a white, edible solid whereas sodium is a metallic solid and chlorine is a green gas, neither of which is edible (and chlorine gas is actually quite dangerous).

> ### Avogadro's number
> Number of atoms or molecules of an element or compound in a mole; equal to 6.022×10^{23}.
>
> ### Mole
> 6.022×10^{23} atoms or molecules of an element or compound.

Common table salt: sodium chloride

In terms of making solutions, combining table salt with water yields a completely different result than mixing either sodium or chlorine with water separately. Making solutions is a fundamental skill for people working in biotechnology. Basic recipes, or **protocols**, can be found in manuals such as <u>Molecular Cloning</u> by Sambrook and Russell (Cold Spring Harbor Laboratory Press).

A common format for protocols includes sections with precise descriptions of reagents (chemicals), supplies, and methods. Here is an example of a common protocol for Luria-Bertani (LB) Broth, which is a nutrient-rich broth used for growing bacteria in the laboratory and is relatively easy to prepare. In this protocol, the pH of the solution is adjusted to 7.5 using sodium hydroxide. In addition, it is sterilized after production by autoclaving, which exposes the solution to high heat under pressure to kill any microorganisms that might be present.

> Add the following to 800ml H_2O:
>
> 10g Bacto-tryptone
>
> 5g yeast extract
>
> 10g NaCl
>
> Adjust pH to 7.5 with NaOH.
>
> Adjust volume to 1L with dH_2O (distilled water).
>
> Sterilize by autoclaving.

Protocols

Procedure or recipe or making a solution or performing a particular technique; guidelines for making or using solutions and equipment in various types of experiments.

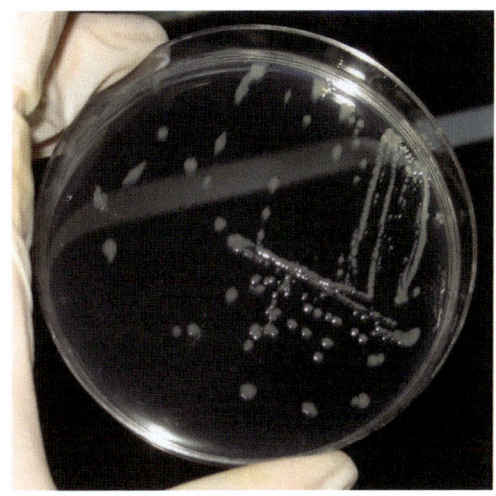

Agar plate prepared with LB broth in a petri dish with bacteria streaked onto it

Acid (acidic)

Having pH less than 7.0.

Base (basic) or Alkaline

Having pH greater than 7.0.

Buffer

Solution that can resist change in pH and maintain a neutral or near neutral pH when strong acids or bases are added.

Molarity

A measure of how much of a compound, in moles/liter, is present in a solution.

Normality

A measure of the acidity or alkalinity of a particular compound in solution.

pH: "potential of hydrogen"

A measure of a substance's alkalinity or acidity; pH greater than 7.0 is basic, while pH less than 7.0 is acidic.

Often, calculations like those we discussed in the first part of this section are necessary in order to complete the process of making a solution. You must be able to measure each component of your solution accurately using the correct measure of weight or mass. You have to understand both **molarity** (M) and **normality** (N). Molarity is the number of molecules of a substance in solution, generally given in moles per liter. For example, a one molar (1M) solution contains one mole of substance (6.022×10^{23} molecules) in one liter of liquid. Normality refers to the acid or base concentration of a compound like sulfuric acid (H_2SO_4), which has two hydrogen ions (H+) and thus each mole of sulfuric acid (1M) will have two moles of this acid and be 2 Normal (2N). Another example is sodium hydroxide (NaOH). A 1M solution would contain one mole of hydroxide ions (OH-) that would correlate to one normal of base as well (and be 1N). .

Knowledge of **acids** and **bases** is the key to understanding **buffering**. Buffering is the ability of a solution to resist changes in pH, even when acids or bases are added. In general, most buffers are able to maintain a relatively constant pH at a characteristic value, which will vary between different chemical solutions. How do you know what is an acid and what is a base? Robert Boyle first described acids and bases in the 17th century:

Acids: taste sour, are corrosive to metals, and become less acidic when mixed with bases.

Bases (or alkalis): feel slippery and become less basic when mixed with acids.

It was not until the 1800s that anyone discovered why acids and bases had the properties Boyle described. Svante Arrhenius, a Swedish scientist, suggested that when hydrogen ions (H+) are disassociated (removed) from their compounds in water, acids are formed; whereas when hydroxide ions (OH-) are disassociated with their compounds, bases are formed. Neutralization occurs when an acid and a base are mixed together to produce water and salt. For example, mixing hydrochloric acid (HCl) with sodium hydroxide (NaOH) to produce water (H_2O) and table salt (sodium chloride; NaCl).

Acids and bases. Lemon juice is an acid and laundry detergent is a base.

pH is a measure of a substance's acidic or basic nature. If a substance has a pH less than 7, it is considered an acid. If a substance has a pH greater than 7, it is considered a base. A substance with a pH of exactly 7 is considered neutral. Weak acids and bases are useful in biotechnology because of their ability to **buffer** solutions, since most living organisms rely on very small pH fluctuations (for example blood plasma can fluctuate between 7.35 and 7.45; a very small amount) in order for us to survive. Many enzymes need particular pH conditions in order to work. Buffer solutions, therefore, are solutions that can resist changes in pH and maintain a neutral or near neutral pH when strong acids or bases are added. Some common buffers used in biotechnology include Tris, Hepes, PIPES, MOPS, and others.

Summary

Acid and base chemistry, along with the phenomenon of buffering are important concepts for the biotechnologists to understand. Maintaining conditions in the laboratory that approximate the conditions and chemistry inside living cells are often important for success in the industry.

Concentration of Hydrogen ions compared to distilled water			Examples of solutions and their respective pH
1/10,000,000	14	Liquid drain cleaner, Caustic soda	
1/1,000,000	13	bleaches, oven cleaner	
1/100,000	12	Soapy water	
1/10,000	11	Household Ammonia (11.9)	
1/1,000	10	Milk of magnesium (10.5)	
1/100	9	Toothpaste (9.9)	
1/10	8	Baking soda (8.4), Seawater, Eggs	
0	7	"Pure" water (7)	
10	6	Urine (6) Milk (6.6)	
100	5	Acid rain (5.6) Black coffee (5)	
1,000	4	Tomato juice (4.1)	
10,000	3	Grapefruit & Orange juice, Soft drink	
100,000	2	Lemon juice (2.3) Vinegar (2.9)	
1,000,000	1	Hydrochloric acid secreted from the stomach lining (1)	
10,000,000	0	Battery Acid	

Table illustrating the pH of common solutions

The ability to understand and prepare chemical solutions accurately is a key to biotechnology, as well as other scientific disciplines, and often relies on understanding the periodic table of elements.

Concept Reinforcement

1. Work out the molecular weights of each of the following molecules or compounds. Use the periodic table to help you.

    a. Fe_2O_3

    b. $Al(NO_3)_3$

    c. $Al_2(SO_4)_3$

2. If iron has a molar mass of 55.8 grams/mole, how many molecules are in 55.8 grams of iron?

3. Molarity versus Normality: Given 1 mole of the compounds below in a one liter solution, what are the molarity and normality of each?

    a. HCl

    b. Na_2HPO_4

    c. H_2SO_4

    d. $(NH_4)OH$

Section 2.2 – SI System of Measurement; Making Dilutions

Section Objectives

- Describe the SI system and identify the base units and prefixes used for measurement

- Describe how dilutions are created from solutions

International System of Units and the Metric System

The **International System of Units** (abbreviated **SI** from the French Le Systeme International d'Unites) is the modern form of the metric system that is used most widely in the world of science and commerce. It is based on the number 10, primarily because of the convenience of mathematical manipulations with this number and its multiples. It is based on the older metric system, which encompassed several units (centimeter-gram-second), but the SI includes meter-kilogram-second instead. In addition, the SI introduced several newly named units and is a dynamic system, in which units are created, and definitions are modified through international agreement.

> **International System of Units (SI)**
>
> The modern form of the metric system.

The original metric system was conceived by a group of scientists who were commissioned by King Louis XVI of France to create a unified and rational system of measures in the late 1700s. It was adopted by France soon thereafter, but was not recognized worldwide until after World War II. In 1960, it was named the International System of Quantities (ISQ). The power of the metric system is the use of a single unit of measure for any physical quality. Since all measurements are all based on the number 10 and utilize decimal places and exponents, calculations between measurements are relatively easy.

The SI system consists of a set of units together with a set of prefixes. The SI units can be divided into two subsets, those above (larger than) the base unit and those below (smaller than) the base unit. There are seven base units that allow for measurement of the physical properties of length, mass, time, electric current, thermodynamic temperature, amount of a substance, and luminous intensity (see table). An eighth base unit for volume, the liter, is often included; it is actually an extension of length. These base units are then modified with a prefix to indicate which integer multiple of 10 the measurement is. The original base units were defined by physical characteristics of natural objects such as Earth and water. For example, the gram was defined as the weight of 1 milliliter of water near its melting point.

SI base units

Name	Symbol	Quantity
Meter	M	Length (1 meter = the distance light travels in an absolute vacuum during 1/299,792,458 of a second)
Kilogram	Kg	Mass (1 Kg = the mass of a platinum-iridium cylinder that is kept in a vault in France)
Liter	L	Volume (1000L = 1 cubic meter; 1L = 1 cubic decimeter)
Second	S	Time (1 second = 9,192,631,770 periods of vibration of the radiation emitted at a specific wavelength by an atom of Cesium-133)
Ampere	A	Electric current
Kelvin	K	Thermodynamic temperature
Mole	Mol	Amount of a substance
candela	c	Luminous intensity

SI prefixes

Name	Symbol	Factor
Yotta-	Y	10^{24}
Zetta-	Z	10^{21}
Exa-	E	10^{18}
Peta-	P	10^{15}
Tera-	T	10^{12}
Giga-	G	10^{9}
Mega-	M	10^{6}
Kilo-	K	10^{3}
Hecto-	H	10^{2}
Deca-	Da	10^{1} (10)
(Base Unit)		10^{0} (1)
Deci-	d	10^{-1}
Centi-	c	10^{-2}
Milli-	m	10^{-3}
Micro-	u	10^{-6}
Nano-	n	10^{-9}
Pico-	p	10^{-12}
Femto-	f	10^{-15}
Atto-	a	10^{-18}
Zepto-	z	10^{-21}
Yocto-	y	10^{-24}

Many units used in everyday life and in science are not derived from the seven base SI units combined with the SI prefixes. In some cases, these deviations have been approved by the **BIPM** (Bureau International des Poids et Mesures). Examples include minutes/hour/day used for time, Celsius for temperature, the nautical mile and knot for measuring travel distance and the speed of ships and aircraft, astronomical distances measured in light years, blood pressure measured in mmHg (millimeters of Mercury), and atomic mass units (amu) for atomic scale units.

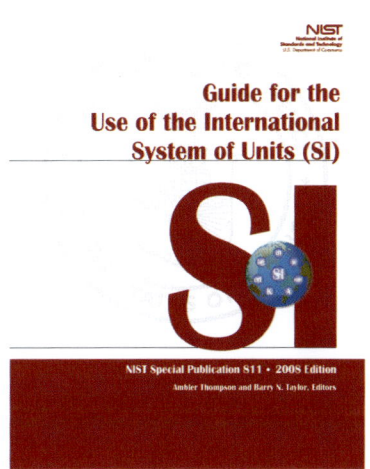

Brochure for the International System of Units

Molarity and Dilutions

In review, **molarity** defines the amount of a substance in moles per liter volume. A **mole** of an element or compound is the number of grams in its **molecular weight** and is equivalent to 6.022×10^{23} molecules of that element or compound. For example, a 1M solution of sodium chloride (NaCl; molecular weight = 58.44 g/L; refer to Periodic Table for the molecular weight of sodium and chlorine) is prepared by dissolving 58.44g of NaCl (table salt) in 1 liter of water.

Once solutions of defined molarity are prepared, it is common that these solutions must be diluted at a later time to the appropriate concentration for a particular experiment or manufacturing process. Dilutions may be done to a single final concentration or they may be serially (sequentially) diluted to multiple different concentrations. Serial dilutions are done such that each dilution is prepared from the previous dilution. That is, a small amount of one solution is added to a given amount of diluent, and then a small amount of this dilution is added to a given amount of diluents again, making a second more dilute solution. Serial dilutions are often performed such that the final concentrations of the substance are in a geometric progression series, that is either linear (dilutions are done such that each successive solution is ½, ⅓, ¼, etc of the previous dilution) or exponential (logarithmic by factors of 10). For example, to prepare serial dilutions that are consecutively 1/10th as concentrated as the previous dilution you could follow the scheme illustrated in the following figure, by removing 1 milliliter from each flask or vial and placing it into 9 milliliters of water or buffer in the next flask or vial. Scientists prepare and use dilutions often in the course of normal research in the laboratory.

Molarity

A measure of how much of a compound in moles/liter is present in a solution.

Mole

6.022×10^{23} molecules of an element or compound.

Molecular weight

The weight in grams of a mole of an element or compound; always expressed in grams per mole; weight in grams of 6.022×10^{23} atoms/molecules of an element or compound.

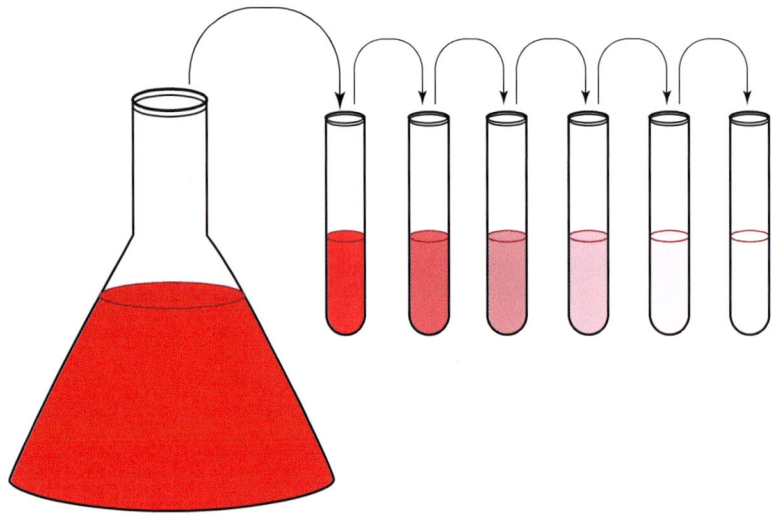

Preparing serial dilutions

Summary

The International System of Units has become the common system of measurements throughout the scientific community worldwide. It allows for the ease of transition between different measurements because of the common set of prefixes. Scientists working in biotechnology use these units for measurement, as well as in making solutions and dilutions, for everyday research and development.

Concept Reinforcement

1. Describe the metric system and the International System of Units. How does the SI differ from the original metric system?

2. How are the base units and prefixes used in the SI?

3. To prepare a 0.1M solution (which is equivalent to 100millimolar (mM), how much of the following compounds would you have to add to 1 liter of water? Use the periodic chart in Section 1 to determine the molecular weight of each compound first.

    a. $MgSO_4$

    b. NaOH

    c. KCl

4. How would you prepare 100milliliters (mls) of a 1M solution of KCl?

Section 2.3 – Limits of Measurement

Section Objective

- Discuss some of the limitations of measurement and the terms used to describe them

Limits of Measurement

Once an experiment has been designed, most often various types of measurements are taken to determine the effect of one variable on another. These measurements will then be used to test the hypothesis of the experiment. The measurements taken can be assessed for their usefulness and robustness (quality) by determining both their accuracy and precision. The **accuracy** of any measurement is an indication of how close the resulting experimental value is to an established reference standard for that technique and sample type. Reference standards are samples for a given type of measurement that have a known output or signal. They are used as a type of positive control in many different types of experiments and assays. Many established reference standards are available from the National Institute for Standards and Technology (**NIST**) in the United States government, but often standards are not available for all techniques or measurements taken in science or biotechnology because they are too new. The **precision** of a measurement is an indication of how closely the individual and replicate (repeated) measurements of the same sample are to each other. The precision of any set of measurements is usually assessed by calculating the variation of each measurement from the **mean** (average) for the replicates. **Standard deviation** and the **coefficient of variance** (CV) are the most common measures of precision. Standard deviation and coefficient of variance can not be calculated for only two replicates, so minimally three replicates (triplicates) are needed to calculate the variation around replicate measurements.

> NIST National Institute for Standards and Technology
>
> A non-regulatory agency of the US Department of Commerce whose mission is to promote US innovation and industrial competitiveness by advancing measurement science, standards, and technology; supplies academia, industry, government, and other users with over 1300 standard reference materials.

For example, let us consider the following experiment. The amount of DNA present in a particular sample was measured using a specific machine and method a total of five times. The results for the five measurements (replicates) were: 100 nanograms/microliter (ng/ul), 98 ng/ul, 105 ng/ul, 110 ng/ul, and 95 ng/ul. The **average** measurement would be: [100 + 98 + 105 + 110 + 95]/5 = 508/5 = **101.6** ng/ul.

Using the formula in the following table, the standard deviation for these five measurements can be calculated. The difference between the five different measurements and the average is: (100 – 101.6 = **-1.6**), (98 – 101.6 = **-3.6**), (105 – 101.6 = **3.4**), (110 – 101.6 = **8.4**), and (95 – 101.6 = **-6.6**). Taking the square of each of these differences will yield: (-1.6 x – 1.6 = **2.56**), (-3.6 x -3.6 = **12.96**), (3.4 x 3.4 = **11.56**), (8.4 x 8.4 = **70.56**), and (-6.6 x -6.6 = **43.56**). The sum of the squares of the differences between the measured values and the average would be: 2.56 + 12.96 + 11.56 + 70.56 + 43.56 = **141.2**. Next you divide this value by the number of total samples minus one: 141.2 / (5 – 1) = 141.2 / 4 = **35.3**. The final calculation to determine standard deviation is to take the square root of this final value: the square root of 35.3 = **5.94**. Thus, for this experiment the average plus or minus the standard deviation would be: **101.6 +/- 5.94 ng/ul**. Thus, the concentration of DNA in the sample is reliably between 95.66 and 107.54 ng/ul.

Term	Definition
Accuracy	How close a measurement falls to a given established standard, often obtained from NIST.
Precision	How close a series of measurements are to each other.
Mean	The average of replicate samples; calculated by summing all of the measurements for a particular sample and then dividing by the number of measurements.
Standard deviation	A common measure of statistical dispersion or variation around repeated measurements of the same sample; that is how close replicate measurements from a single sample are to each other; calculated by taking the square root of the differences between the individual values and the mean squared, divided by the number of replicates. $$\sigma = \sqrt{\frac{\sum (x - \overline{x})^2}{n - 1}}$$ σ = lower case sigma $\sum$ = capital sigma $\overline{x}$ = x bar
Coefficient of Variance (CV)	The ratio of the standard deviation to the mean or average (divide standard deviation by the mean); useful to look at the variation of replicate measurements; often expressed as percent CV. CV = standard deviation / mean

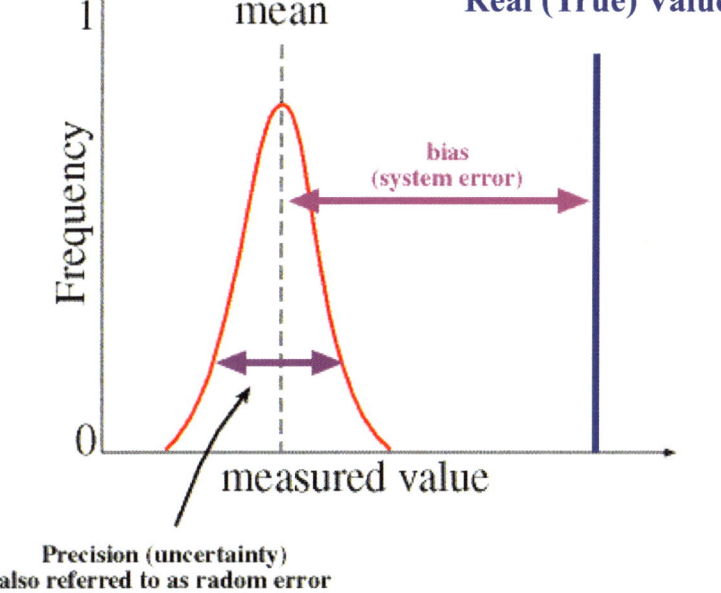

Graphical illustration showing mean (average) and the error if many repeated measurements are taken with the same sample. The curve usually looks like a bell, because most values will be close to the average (height of the bell), but some will be much higher and some much lower (width of the bell). The system error is simply the error that occurs when taking any measurement, as the actual value cannot be determined due to limitations in equipment, etc. The system error may be larger or small depending on the accuracy of the equipment.

Other Limitations to Measurement

Three other terms are often used when referring to the robustness of scientific measurements and experimental techniques: 1) **limit of detection** (L_D or LOD), 2) **limit of quantitation** (L_Q or LOQ), and 3) **minimal detectable amount** (MDA).

The **limit of detection** is defined as the lower limit of detection of a particular substance using a specific technique, and is the lowest amount of a substance that can be distinguished from the absence of the substance. The limit of detection is estimated from the average of the blank measurements (background) and the standard deviation from the average (variation of all of the blank measurements taken). It basically estimates how much of a substance must be present to obtain a reliable measurement that is above the measurement obtained when no substance is present. Generally any measurement that is three standard deviations above the blank value is considered the limit of detection for that substance and technique.

Term	Definition
Limit of Detection	The smallest amount of a substance that can be simply detected by a specific measurement method.
Limit of Quantitation	The smallest amount of a substance that can be accurately quantitated by a specific measurement method.
Minimal Detectable Amount	The smallest amount of a substance that can be reliably detected by a measurement process.
Practical Quantitation Limit	Five times the minimal detectable amount for any measurement process.

The **limit of quantitation** is defined as the value at which the measurement is accurate enough to be quantitative (accurate amount measured) and not just detectable. This value is generally 10 standard deviations above the blank for a particular substance and technique. The **minimal detectable amount** is the amount of substance that can be reliably detected by a measurement process. This value is generally six standard deviations above the blank. The minimal detectable amount is thus between the limit of detection and the limit of quantitation. The **practical quantitation limit** (PQL) is often used because MDA can vary between laboratories, so PQL is defined as 5 times the MDA.

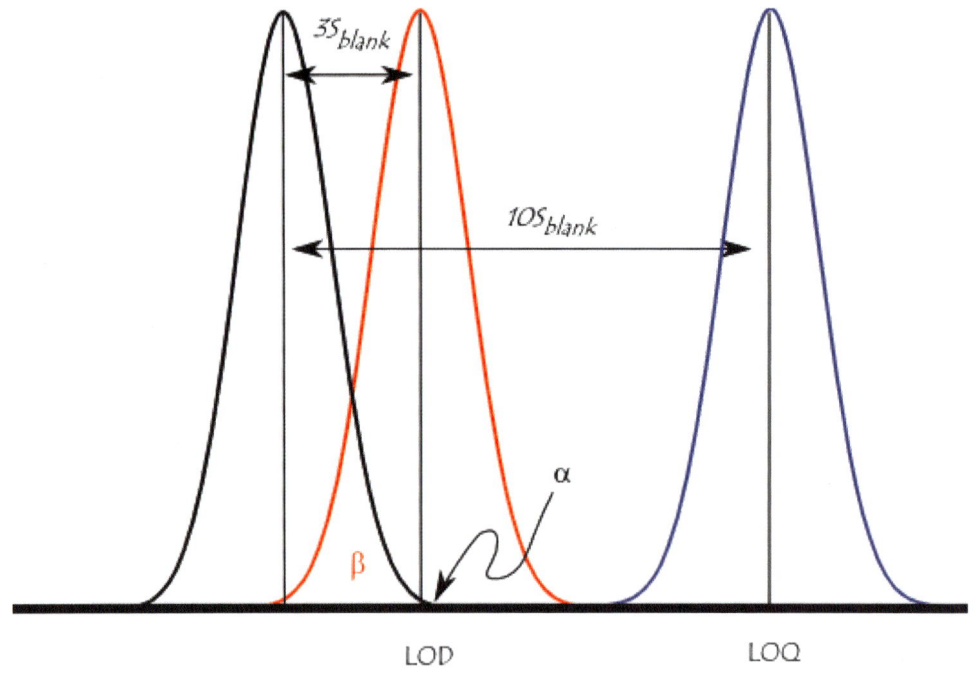

Graphical illustration showing the relationship between LOD and LOQ

To highlight these concepts, consider a situation at a noisy rock concert. If the person next to you tries to speak, but speaks only softly, you will probably not be able to hear him. His voice volume is below the limit of detection. If he speaks a bit louder, you may hear him but it is possible you might not. This represents the limit of detection. If he speaks loud enough that you can always hear him, but have trouble understanding what he is saying, that represents the minimal detectable amount. If however, he speaks louder, you can then not only hear him, but understand him as well. This represents the limit of quantitation.

Thus although many different types of substances, activities, and signals are measured using many different techniques in biotechnology, they all have limits to their ability to detect the desired substance and this information is critical when interpreting experiments that utilize these techniques.

Summary

Experiments are designed to investigate the relationship between different factors in an environment. In order to detect various relationships, many different types of measurements are taken, and then conclusions drawn from the results. Scientists need to be aware that within all measurements errors can occur and that there are limitations to such measurements. Various types of calculations are then used to verify the measurements can be interpreted with confidence.

> ### Detectable amounts.
>
> A technique often used in biotechnology is the quantitation of the amount of nucleic acid (DNA or RNA) present in a particular sample. This is done by measuring the absorbance of the material at 260 nanometers of light. Nucleic acids, such as DNA and RNA, absorb light at this particular wavelength, and the amount of light absorbed is proportional to the amount of DNA or RNA present in the sample. Thus, the absorbance measurement can be used to determine the amount of DNA present in a liquid sample, and its concentration. This information is very useful and often required for further analysis or manipulation of the DNA. The measurements require a special machine called a spectrophotometer. A particular model of spectrophotometer can detect DNA when the concentration is at least 2nanogram/microliter (ng/ul). This can be called the limit of detection for this particular sample type and machine. It can consistently and accurately detect DNA when the concentration is at least 50ng/ul., which can be considered the limit of quantitation. The lowest amount of DNA that can be detected for this machine is 15ng/ul and is considered the minimal detectable amount. Notice the minimal detectable amount lies between the limit of detection and the limit of quantitation.

Concept Reinforcement

1. What is limit of detection and how might it be used in an experiment?

2. What is limit of quantitation and how might it be used in an experiment?

3. What is minimal detectable amount and how might it be used in an experiment?

4. What is practical quantitation limit and how might it be used in an experiment?

Section 2.4 – The Eukaryotic Cell

Section Objectives

• Describe the major categories of eukaryotic organisms

• Describe the major features of the various organelles in a eukaryotic cell

Eukaryotic Organisms and the Phylogenetic Tree

Organisms in general can be divided into three main divisions: 1) **prokaryotes**, 2) **archaea**, and 3) **eukaryotes**. The prokaryotic organisms, which consist primarily of bacteria, are structurally similar to the archae. The predominant distinction between these two divisions and eukaryotic organisms is the presence or absence of a nucleus. Both bacteria and archaea lack a defined nucleus. However, the archae are much more similar to eukaryotic organisms in their enzymatic processes for making RNA and proteins, molecules necessary for life

Eukaryotes are believed to have evolved approximately 1.6-2.1 billion years ago, while prokaryotes are believed to have evolved about 3.5 billion years ago.

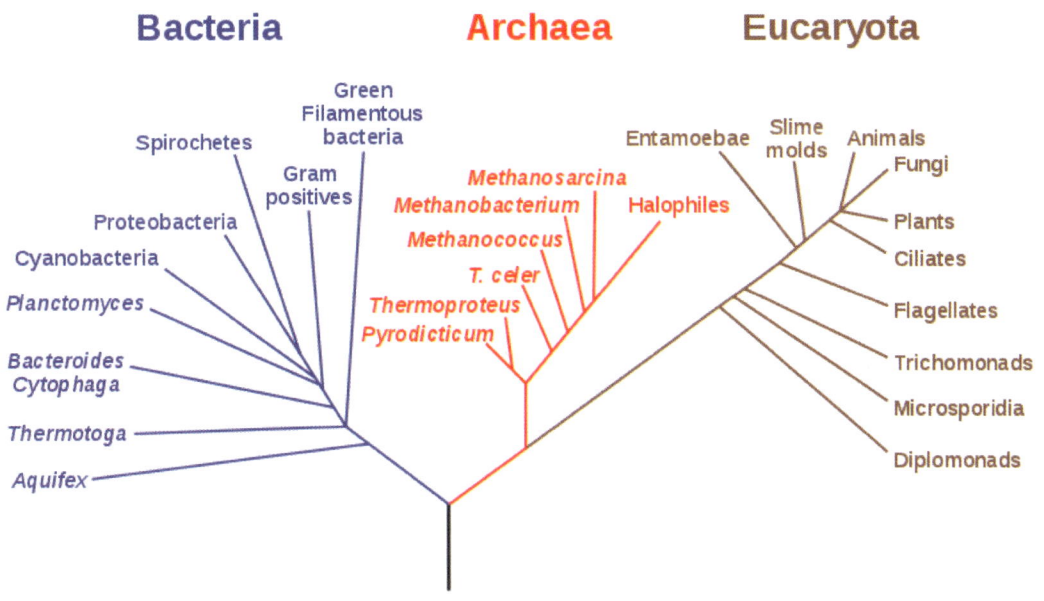

Phylogenetic tree of life showing relationship between prokaryotes (bacteria), archae, and eukaryotes

Prokaryote
A cell or organism that does not contain a defined nucleus.

Archaea
Archaea were originally named archaebacteria, because of their structural similarity to prokaryotic bacteria, and renamed after the realization that they are actually more similar to eukaryotes; they may be the oldest organisms on earth and tend to inhabit very extreme environments.

Eukaryote
A cell or organism that contains a defined nucleus.

Eukaryotic organisms consist of animals (including vertebrates, invertebrates, insects, etc.), plants, fungi, and **protists**. All eukaryotes contain complex structures that are organized in membrane-bound **organelles**. Protists are a diverse group of eukaryotic organisms that all have relatively simple organization, usually being either unicellular (one-celled) or multicellular without highly specialized tissues. This broad group contains the one-celled animal-like protozoa (flagellates, amoeboids, ciliates, and sporozoa), the plant-like protophyta (chlorophytes, rhodophytes, heterokontophytes), and the fungus-like slime and water molds. Some of the most common members of the protist group include *Euglena*, *Amoeba*, *Paramecium*, *Toxoplasma*, diatoms, and red, green, and brown algae.

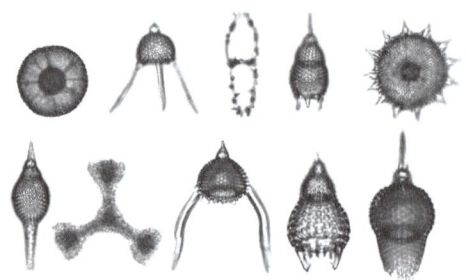

Various eukaryotic protist organisms

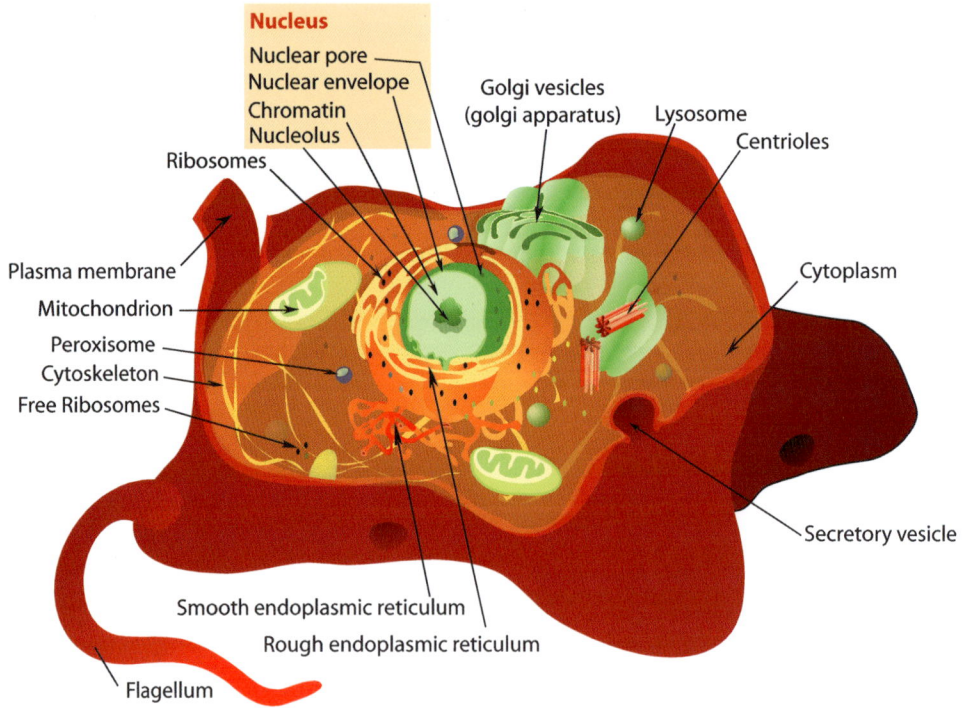

Diagram illustrating the different organelles in a eukaryotic cell

Characteristics of Eukaryotic Organisms

The characteristic feature of a eukaryotic organism that differentiates it from prokaryotic or archaea organisms is the presence of a defined nucleus. The presence of the nucleus gives the organisms their name, which comes from the Greek work ευ, meaning "good/true", and the Greek work καρυον, meaning "nut." The nucleus contains the cellular genetic information (DNA), which is organized into long, linear complexes with proteins to form chromosomes. The function of the nucleus is to maintain the integrity of the genetic information and to control the expression of the genes into RNA. It was the first organelle discovered by Franz Bauer in 1802, and later described in more detail by Robert Brown in 1831. It occupies about 10% of the total volume of the cell, but may be higher in some cell types.

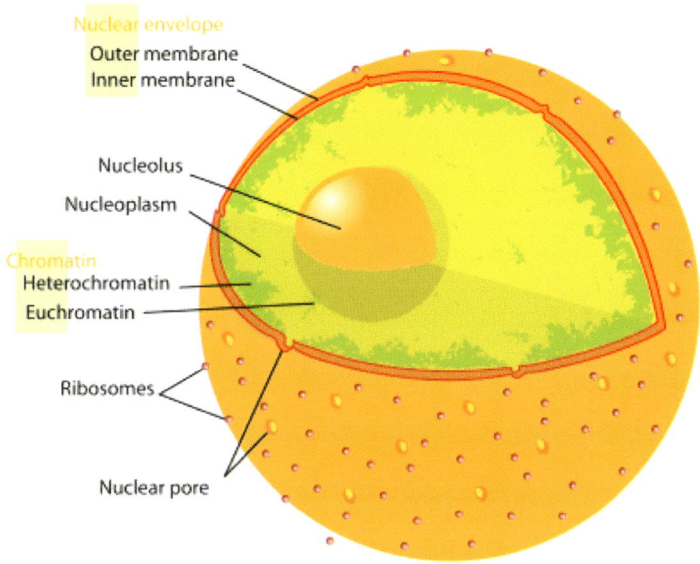

Detailed structure of the eukaryotic nucleus. The nucleus can be divided into compartments, including the nuclear membranes (inner and outer), nucleolus, and nucleoplasm.

The main structure of the nucleus is the nuclear envelope, a double membrane that encloses the entire organelle and separates it from the surrounding **cytoplasm**. Inside the nucleus is the nuclear lamina, a meshwork of proteins that add mechanical support to the nucleus, similar to the **cytoskeleton** that supports the cytoplasm and the cell. The nuclear membrane contains many nuclear pores or channels, which allow the movement of specific molecules into and out of the nucleus. The nucleus in general is impermeable to most molecules, so the presence of the nuclear pores is key to allowing various compounds to enter or leave the nucleus. Proteins called karyopherins mediate the movement of larger molecules into and out of the nucleus, such as most proteins, ribosomes, and some RNAs. Nuclear sub-compartments exist, which contain unique proteins, RNA molecules, and particular parts of chromosomes. The nucleolus is one such sub-compartment in the nucleus, and it is the location where **ribosomes** are assembled.

Nuclear compartments

The nuclei of eukaryotic cells contain many different sub-compartments or structures that may play key roles in the function of the nucleus. In addition to the nuclear membrane, nuclear pores, and the nucleolus, other structures have been found. They include Cajal bodies, PML bodies, and speckles and paraspeckles to name a few. Cajal bodies are compact structures and a typical nucleus will contain 1-10. They resemble balls of tangled thread under the electron microscope and may be involved in RNA processing. PML bodies are spherical and found scattered throughout the nucleus. They are often associated with the Cajal bodies and may regulate RNA production.

Cytoplasm

A semi-transparent fluid that fills cells; makes up about 70% of the cell and is composed of water, salts, and organic molecules; it is the site where most cellular activities are performed.

Cytoskeleton

The supporting framework or scaffold within the cell that is composed of various sized filaments; facilitates many functions and activities in the cell, including cellular motion, intracellular transport, and cell division.

Ribosome

A large protein and RNA complex that cooperates to perform translation (protein synthesis).

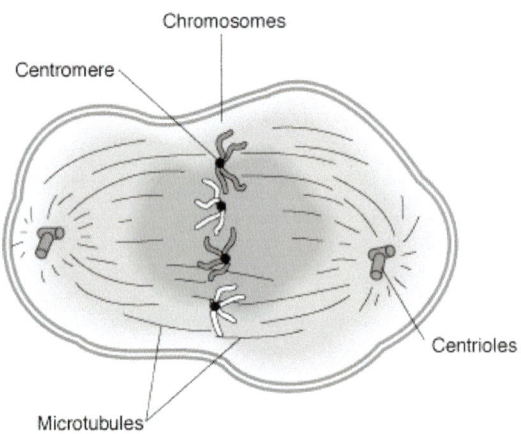

Interphase and metaphase nuclei. In the interphase nucleus the DNA is much looser in its conformation, while in the metaphase nucleus the DNA has replicated (copied itself) and condensed into individual chromosomes that can then separate as the cell divides into two daughter cells.

Although most eukaryotic organisms contain a single nucleus, some specialized cell types contain either no nucleus (enucleated) or many nuclei (polynucleated). The best example of an enucleated cell is a mature red blood cell, which loses it nucleus, as well as other organelles, during maturation. Examples of polynucleated cells include some species of protozoa and some fungi, as well as skeletal muscle cells.

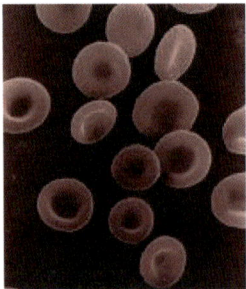

Enucleated red blood cells

102

Other organelles besides the nucleus

Besides the nucleus, eukaryotic cells contain many other types of organelles with complex and vital functions (they are summarized in the table that follows). Various tube- and sheet-like structures extend from the outer nuclear membrane to form the **endoplasmic reticulum** (ER). The ER can be smooth or rough, with the major difference being that the rough ER (RER) is covered in ribosomes. Protein synthesis occurs in the lumen or interior space of the RER, as does protein folding, maturation, and some modifications (decorations) that may be added. The proteins then generally transit to the smooth ER and eventually move in vesicles to the **Golgi Apparatus** (or Golgi Complex) for final processing and release from the cell, incorporation into the cell membrane, or incorporation into **lysosomes**. The primary function of the Golgi Apparatus is to process and package macromolecules such as proteins, lipids, and **proteoglycans**, which are synthesized in the RER and ER. It consists of a series of stacked membranes known as **cisternae** that can be divided into five main regions. Each region contains different enzymes which selectively modify proteins and lipids depending on their final destination. One very important type of modification that can occur to proteins in the RER/ER system and Golgi Apparatus is the addition of sugar (carbohydrate) molecules. Cytoplasmic proteins are synthesized on free ribosomes in the cytoplasm and do not enter the RER system.

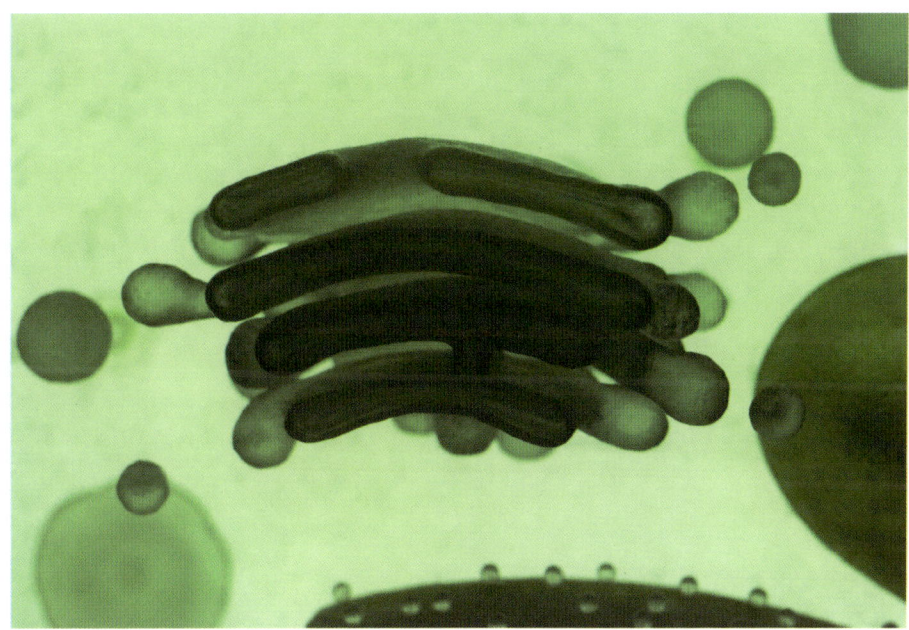

The Golgi Apparatus in a cell, showing the stacked membrane appearance

Mitochondria are found in nearly all eukaryotic cells. They are surrounded by a double-layer membrane, with the inner membrane folded into invaginations called cisternae. The cisternae are the site of aerobic respiration, in which energy from metabolism is converted to chemical energy in the form of **ATP**, with the resulting byproducts being carbon dioxide and water. This process, called **aerobic respiration**, requires oxygen, and is the reason we humans require oxygen to survive. Mitochondria contain their own DNA and can produce the RNA and resulting proteins required for energy production. They divide by **fission** from other mitochondria. Organisms contain numerous mitochondria per cell and in humans and other mammals all mitochondria come from the mother.

Endoplasmic Reticulum

Eukaryotic organelle that forms an interconnected network of tubules, vesicles and cisternae within cells.

Golgi Apparatus

The organelle in eukaryotic cells responsible for protein modification, transport, and release from the cell.

Proteoglycan

The proteins that are highly modified with sugar groups (carbohydrates) that constitute part of the matrix surrounding the outside of cells (extracellular matrix).

Lysosome

The organelle that contains digestive enzymes that are involved in the normal breakdown of cellular components.

Cisternae

The membrane-bound sacks or layers often found in eukaryotic organelles like the Golgi apparatus or mitochondria.

Mitochondria

The organelle responsible for energy production through aerobic respiration. They contain their own DNA and are formed by the fission of other mitochondria.

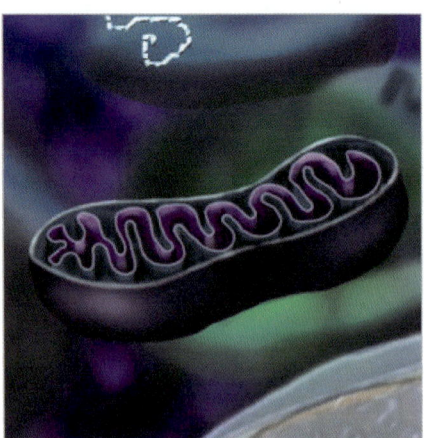

A mitochondria, showing the internal cisternae structure

Other types of organelles are present in eukaryotic cells. **Lysosomes** are membrane-bound compartments involved in the break-down or degradation of larger molecules in the cell. These compartments contain many digestive enzymes that break down proteins and carbohydrate (sugar) complexes. **Vesicles** are membrane-bound sacks that transfer various molecules within the cell or to the outside of the cell for excretion.

All eukaryotic cells are surrounded by a **cell membrane** and internally supported networks of protein and carbohydrate filaments called the **cytoskeleton** or cellular scaffold. The cytoskeleton plays many important functions in the cell, including mechanical support, cellular motion, intracellular transport, cell division, and cellular signaling. Eukaryotic cells contain three types of cytoskeletal filaments, which are called **microfilaments**, **intermediate filaments**, and **microtubules**. Microfilaments are composed of a specific protein called **actin** and are concentrated just below the cell membrane. Intermediate filaments are composed of different proteins depending on the specific cell type and are more stable than microfilaments, as they can resist compression and better maintain cell shape. They are also responsible for anchoring organelles within the cytoplasm. Microtubules are hollow cylindrical fibers that are composed of a protein called **tubulin**. They are primarily involved in intracellular transport and in mitosis during spindle formation. They are also involved in the synthesis of the cell wall in plants.

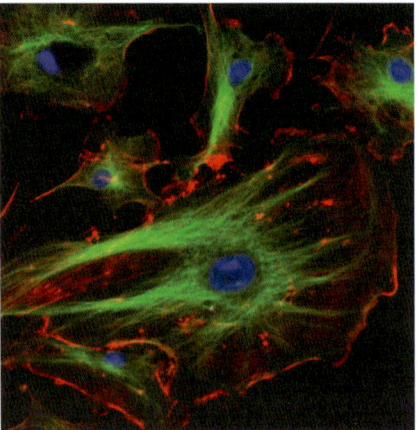

Components of cytoskeleton labeled with fluorescently-tagged antibodies. Nuclei are shown in blue and the cell membrane appears red.

104

Thus, eukaryotic cells and organisms are quite complex and, even within the category of eukaryote, can differ significantly. This is highlighted by the differences in single-celled organisms versus multicellular organisms. Eukaryotic cells have evolved to possess numerous organelles with very specialized functions that allow the cell to produce energy, move, replicate, remove waste products, and communicate with its neighbors.

Major organelles of eukaryotic cells or organisms

Organelle	Main Function	Structure	Organisms	Comments
Cell membrane	Encloses cellular components	Double-membrane compartment	All eukaryotes	
Nucleus	DNA storage and RNA synthesis	Double-membrane compartment	All eukaryotes	Protects genetic information (DNA)
Mitochondria	Energy production (through ATP generation)	Double-membrane compartment	Most eukaryotes	Contains own DNA; theorized to be derived from engulfed bacteria in ancestral eukaryotic cell (endosymbiosis)
Endoplasmic reticulum (ER)	Protein synthesis and protein folding; synthesis of lipids	Single-membrane compartment	All eukaryotes	Rough ER is covered in ribosomes (RER) and folds are sack shaped; smooth ER folds are tubular in shape
Golgi Apparatus (Golgi Complex)	Sorting and modification of proteins	Single-membrane compartment	All eukaryotes	Convex face (cis) is nearest to the RER; concave face (trans) is farthest from the RER
Chloroplast (plastid)	Photosynthesis	Double-membrane compartment	Plants and some protists	Contains own DNA; theorized to be derived from engulfed photosynthetic bacteria in ancestral eukaryotic cell (endosymbiosis)
Vacuole	Storage and homeostasis	Single-membrane compartment	Most eukaryotes	
Lysosome	Breakdown (degradation) of large molecules (like proteins and polysaccharides)	Single-membrane compartment	Most eukaryotes	
Vesicle	Transport of materials inside of cell and may be excreted outside of cell	Single-membrane compartment	All eukaryotes	

Summary

The eukaryotic cell is composed of many different and complex organelles that perform essential functions to the cell, from storing the genetic information, allowing for new proteins to be made, supporting the cell's structure, and generating the energy that the cell needs to grow and divide.

Concept Reinforcement

1. What is the primary difference between eukaryotes, prokaryotes, and archae organisms?

2. What two proteins are involved in the cytoskeleton and what role do they play?

3. What is the main function of mitochondria in the cell?

Section 2.5 – The Prokaryotic Cell

Section Objectives

- Identify what a prokaryotic cell is and discuss its properties

- Discuss archaea organisms and their differences from prokaryotes

Prokaryotic Cells

Prokaryotes are a division of organisms that lack a defined **nucleus**, as well as other membrane-bound organelles such as mitochondria, endoplasmic reticulum, Golgi Apparatus, and lysosomes. Thus the DNA in prokaryotic cells is floating in the cytoplasm. Prokaryotes consist primarily of two different domains: bacteria and **archaea.** However, archae are considered on a separate branch of the phylogenetic tree from both prokaryotes and eukaryotes, though their intracellular enzymatic processes are much more similar to eukaryotes than prokaryotes. The vast majority of prokaryotic cell organisms consist of individual cells, but a few exist as multicellular organisms. Bacteria exist in three basic shapes: cocci (spherical), bacilli (rod shaped), or spiral (curved).

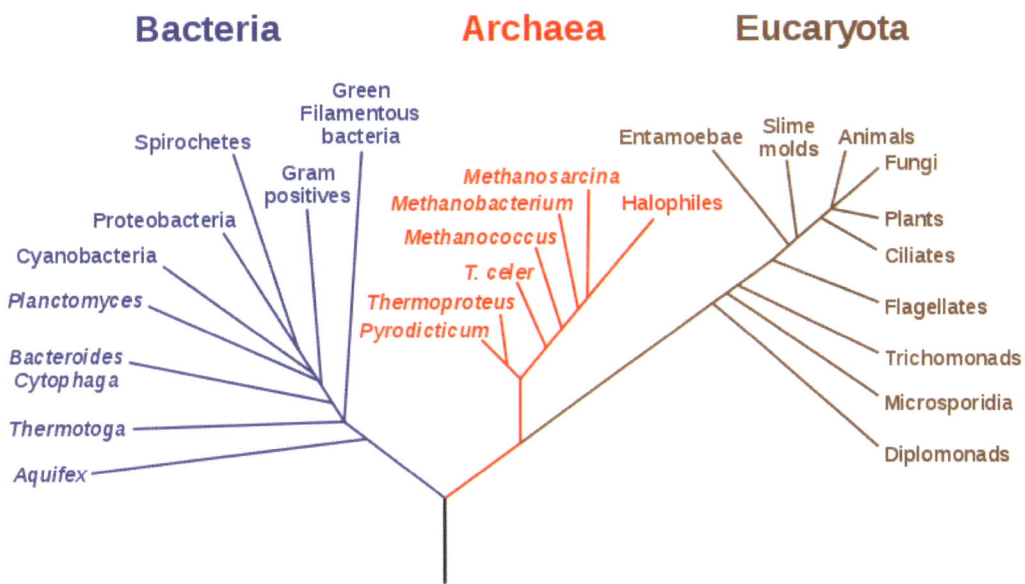

Phylogenetic tree of life showing relationship between prokaryotes (bacteria), archaea, and eukaryotes

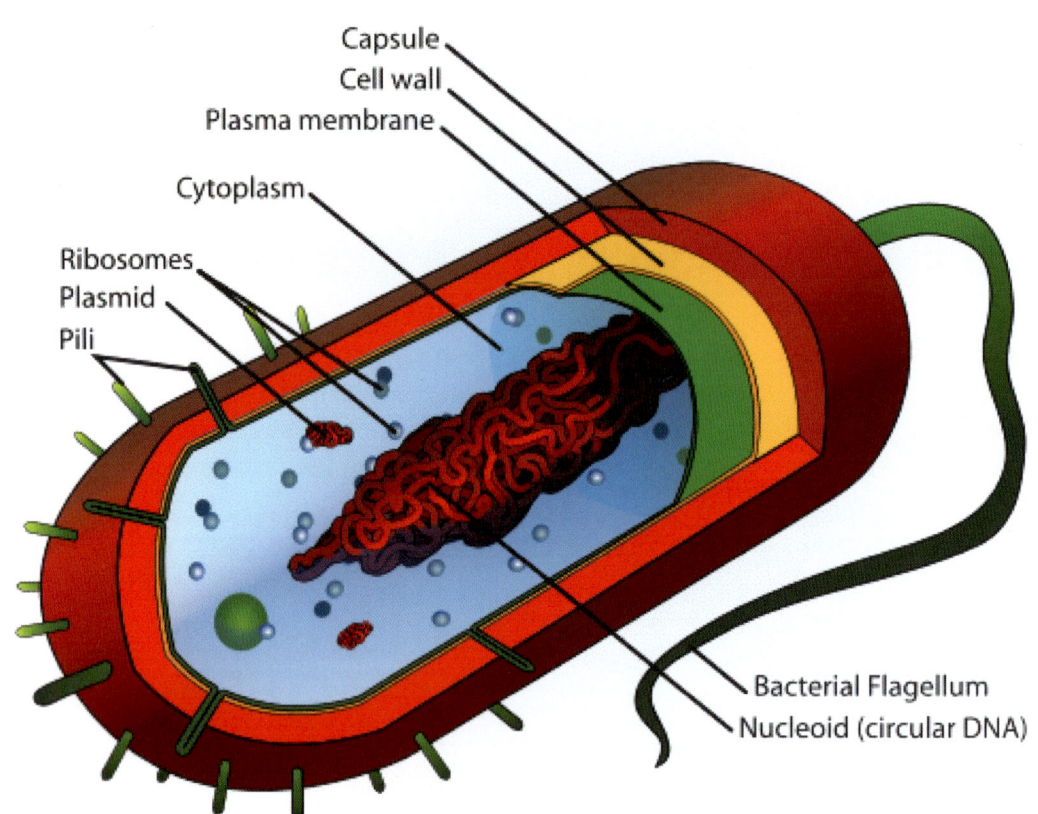

Structure of a prokaryotic bacteria cell showing the various components such as DNA (nucleoid), cytoplasm, ribosomes, cell membrane, and cell wall

Archaea were originally named archaebacteria, but as indicated earlier, they have prokaryotic structural features, they are more closely related to eukaryotic organisms and are thus distinct from bacteria. While archaea are similar to prokaryotic organisms in cell structure and metabolism, their processes of RNA and protein synthesis are most similar to eukaryotes. Recent studies have suggested that archaea may be the most ancient organisms on earth. They tend to inhabit extreme environments, but can be found in all habitats. They can survive extremely high or low temperatures, salty, acidic (low pH), alkaline (high pH), or otherwise harsh environments. Significantly, large numbers of archaea are found in the ocean, accounting for up to 40% of the biomass (biological material) present. The extreme environments that archaea survive in may make them useful targets for biotechnology in the search for enzymes or proteins that can tolerate harsh conditions. In addition, the unusual metabolism they can exhibit (nitrifiers, methanogens, and methane oxidizers) also make them appealing tools for biotechnology applications.

Bacteria contain their genomic DNA in a single loop of chromosomal DNA in the cytoplasm. This single chromosome contains all of the genes necessary for bacterial survival. Bacteria also contain small circular pieces of DNA called **plasmids**, in which they carry genes useful for adaptation to different environments, such as resistance to antibiotics. Bacteria can transfer these plasmids to one another and thus transfer the environmentally adaptive traits. This feature has allowed bacteria to adapt and evolve quickly to take advantage of changing conditions in their environments.

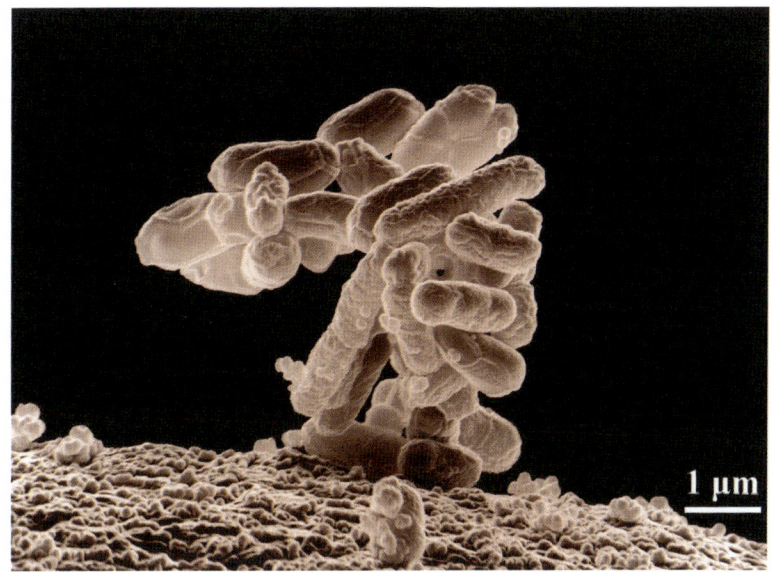

E. Coli bacteria.

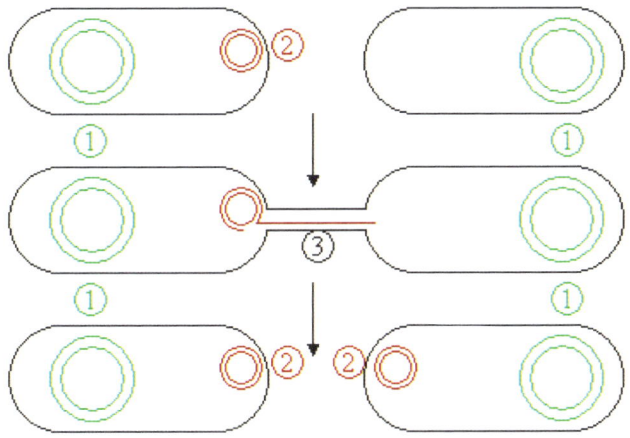

Bacteria containing genomic bacterial DNA and smaller plasmid DNA.
Note how the plasmid DNA can be transferred from one bacterium to another
through a process called conjugation.

Gram- versus Gram+ Bacteria

Bacteria are commonly classified as gram- or gram+ based on their ability to retain the stain crystal violet following a destaining protocol. Gram+ bacteria will retain the dye solution following treatment with the decolorizing solution and appear purple, while gram- bacteria do not. The gram- bacteria actually take up the counterstain and will appear a different color (red or pink) as compared to the gram+ cells. The ability to retain the dye following destaining or counterstaining is due to differences in the cell wall components of these two types of bacteria: gram+ bacteria have much more peptidoglycan in their walls as compared to gram- bacteria. Gram+ bacteria also often lack the extra outer cell membrane that gram- bacteria have. Peptidoglycans are a complex mixture of sugars and amino acids that form a meshwork around gram + bacteria. These peptidoglycans can affect the ability of the bacteria to infect and grow in certain host organisms.

Escherichia coli bacteria

E. coli are gram-negative bacteria that are commonly found in the lower intestines of most animals, including humans. Most strains are harmless and part of our normal flora, though some strains can be pathogenic (i.e. serotype O157:H7). The bacteria were first identified by Theodor Escherich in 1885, who was a German pediatrician and microbiologist. They are easy to grow, and because they are generally not harmful, this particular strain of bacteria has become the most commonly used strain in research and biotechnology. In addition, its genetics are relatively simple and easy to manipulate, also making it an ideal organism for research and as a tool. Cultivated strains have become well adapted to the laboratory environment and become a very efficient host for the production of proteins from a variety of sources, including humans.

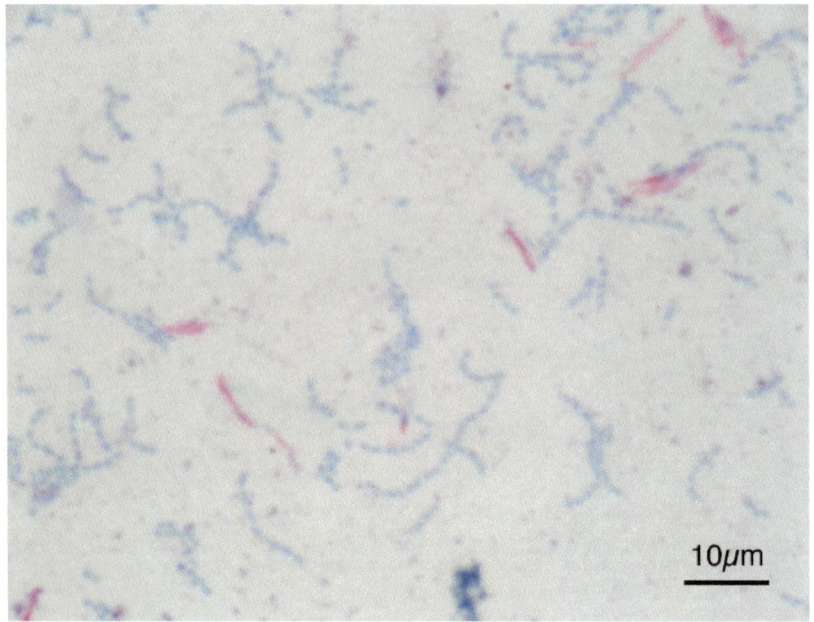

Gram- and gram+ bacteria showing the differential staining patterns.
Gram+ will appear purple/blue while gram- will appear red or pink.

Bacteria can grow in a wide variety of conditions and environments (though generally not as harsh as archaea) and they have a broad range of environment and nutrient requirements. Most bacteria are non pathogenic (do not cause disease) and include the bacteria that normally live in our gastrointestinal tract. However, some bacteria can be pathogenic and can cause a variety of diseases such as food poisoning, cholera, bubonic plague, skin infections, and strep throat. Bacteria also play a key role in the global ecosystem, since they are responsible for the vast majority of nutrient cycling that occurs as living things die and are degraded by bacteria back into their molecular components, to be reused by other organisms.

Summary

Prokaryotic organisms (bacteria) have found a special place in biotechnology and can be considered the foundation for both classical and modern biotechnology. The applications in which they are used include protein production (through fermentation), bioenergy production, toxic waste cleanup (bioremediation), and recombinant DNA production (i.e. plasmid expression).

Concept Reinforcement

1. What is the name for the branch of single-celled organisms that are separate from prokaryotes and eukaryotes and possess similarities with each and what is unique about their growth environments?

2. How could archaea be utilized in biotechnology?

3. What is the most common strain of bacteria used in biotechnology and why was it selected?

4. What is the difference between gram+ and gram- bacteria?

Section 2.6 – Compare and Contrast Mammalian, Plant, and Prokaryotic Cells

Section Objectives

- Describe the similarities and differences between eukaryotic cells and prokaryotic cells

- Describe the similarities and differences between mammalian (animal) cells and plant cells, both of which are eukaryotic

Eukaryotes Versus Prokaryotes

As discussed, organisms can be divided into three main divisions: 1) **prokaryotes**, 2) **archaea**, and 3) **eukaryotes**. The primary difference between prokaryotes and archaea with eukaryotic cells is the presence of a defined nucleus, as well as other membrane-bound organelles. Prokaryotic organisms tend to be single-celled while eukaryotic organisms tend to be multi-cellular and in general much more complex. Eukaryotic cells also have much lower metabolic rates and longer generation (replication or doubling) times as compared to prokaryotic cells. Prokaryotic organisms are generally bacteria, while eukaryotic organisms include single-celled **protists**, yeast, and algae, and multi-cellular plants, **fungi**, and animals (all vertebrates and invertebrates, including mammals, reptiles, amphibians, birds, fish, insects, etc.).

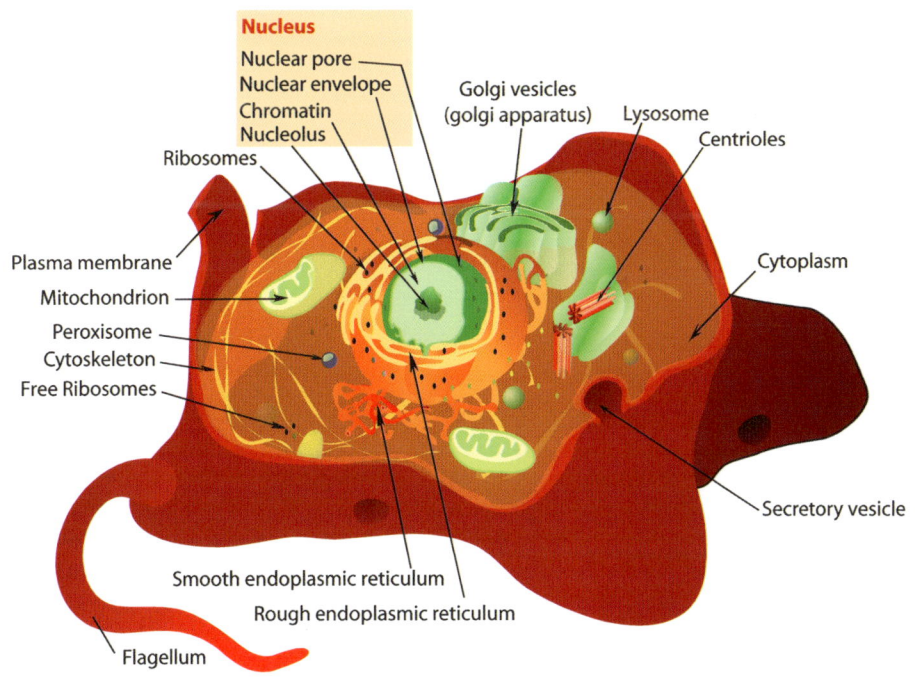

Typical eukaryotic cell showing the nucleus and various other organelles that are characteristic of this type of cell

Prokaryotes

Prokaryote: cell that does not contain a nucleus.

Archaea

Single-celled organisms similar to bacteria, but they evolved separately from them and are probably the oldest organisms on earth. They were originally named archaebacteria. This name was later changed since they are more closely related to eukaryotes.

Eukaryotes

Cells that contains a nucleus.

Protists

A diverse group of eukaryotes that are either single-celled or multi-cellular but lacking specialized tissues.

Fungi

A diverse class of eukaryotic organisms that include molds, mushrooms, and yeast, which are used for many types of fermentation, such as in the production of bread, wine, and beer.

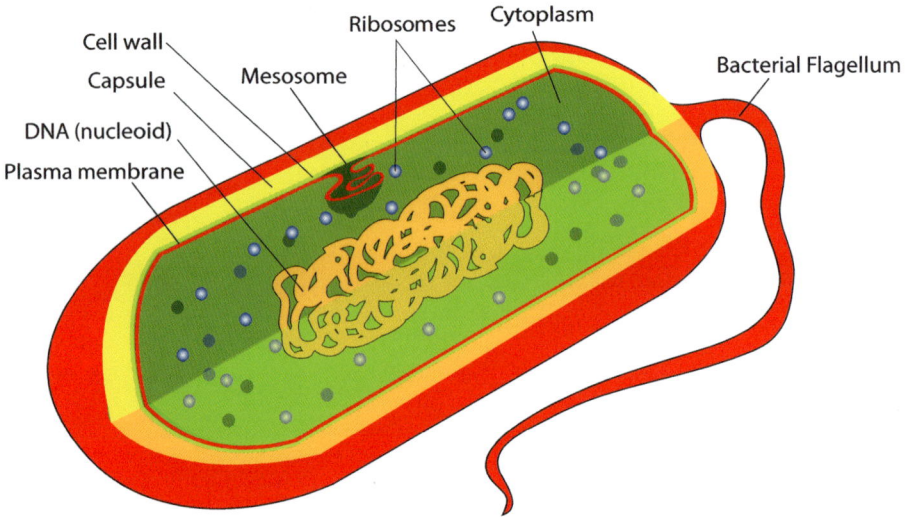

Plasmid

The double-stranded, circular pieces of DNA that are found in bacteria and yeast, and are separate from the genomic or chromosomal DNA. Plasmids often possess genes advantageous for different environmental conditions and can be passed from one bacteria to another.

Intron

The noncoding sections in DNA and then mRNA that are removed in the nucleus in a process of RNA processing called splicing. These sequences do not code for a protein.

Exon

The region of DNA within a gene that is contained within the final messenger RNA and contains the coding information for making a protein.

Typical prokaryotic cell (or bacteria) showing the various molecular components in the cell that are characteristic of this cell type

In contrast to bacteria and archaea, eukaryotic cells contain their genetic material in the nucleus of the cell and often contain multiple different chromosomes with the DNA tightly wound around protein cores, while bacteria contain a single loop of chromosomal DNA that is floating free in the cytoplasm. Bacteria also contain small circular pieces of DNA called **plasmids**, in which they carry genes useful for adaptation to different environments. Eukaryotic cells do not normally possess plasmid DNA.

The presence of the genomic DNA in the nucleus allows eukaryotic cells more control over gene expression (RNA synthesis and thus protein production) than prokaryotic cells, since in bacteria RNA synthesis and protein production occur simultaneously in the cytoplasm. Eukaryotic cells possess DNA sequences called **introns** in their genes, which prokaryotic cells do not have. These intron sequences are removed from the messenger RNA in the nucleus before the mature RNA is exported from the nucleus to the cytoplasm for translation (protein production), where only the **exon** sequences are converted to protein.

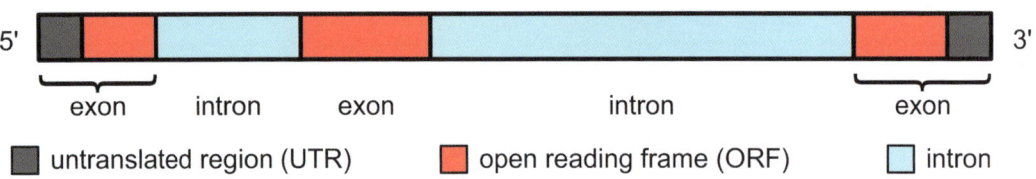

Diagram showing the structure of a typical eukaryotic gene, with introns and exons. The introns are removed from the mRNA before being used as a template to make a protein.

Most eukaryotic cells contain a single nucleus, but some species of protozoa and some fungi contain multiple nuclei (called polynucleated). It has been theorized that eukaryotic cells arose during evolution from the symbiotic relationship between a bacteria and an archaea, in which an archaea invaded and then resided within the bacteria, giving the new cell the characteristics of the archaea. This is similar to the accepted genesis of eukaryotic mitochondria and chloroplasts, as they are also believed to be bacteria that were incorporated into a very early eukaryotic cell. Both of these organelles are enclosed by a cell membrane, just like bacteria.

114

Mammalian (Animal) Cells Versus Plant Cells

While plant cells are also eukaryotic, they are very different from animal cells. They contain a cell wall composed of cellulose, proteins, and lignins that provide a very tough outer support for plant cells. Plants and various types of algae also contain organelles called plastids (also called chloroplasts). Plastids possess their own DNA similar to mitochondria, but have the added function of containing chlorophyll that enables energy production by photosynthesis. Mitochondria are the membrane-bound organelles within eukaryotic cells that are responsible for energy generation. Other types of plastids are involved in the storage of food. The source of plastids/chloroplasts is believed to be through the symbiotic relationship between a eukaryotic cell and a cyanobacteria that is capable of photosynthesis. That is, sometime early in the evolution of plant cells, a simple eukaryotic cell engulfed a bacteria that is capable of photosynthesis (cyanobacteria), thus giving the eukaryotic cell the ability to produce energy from light.

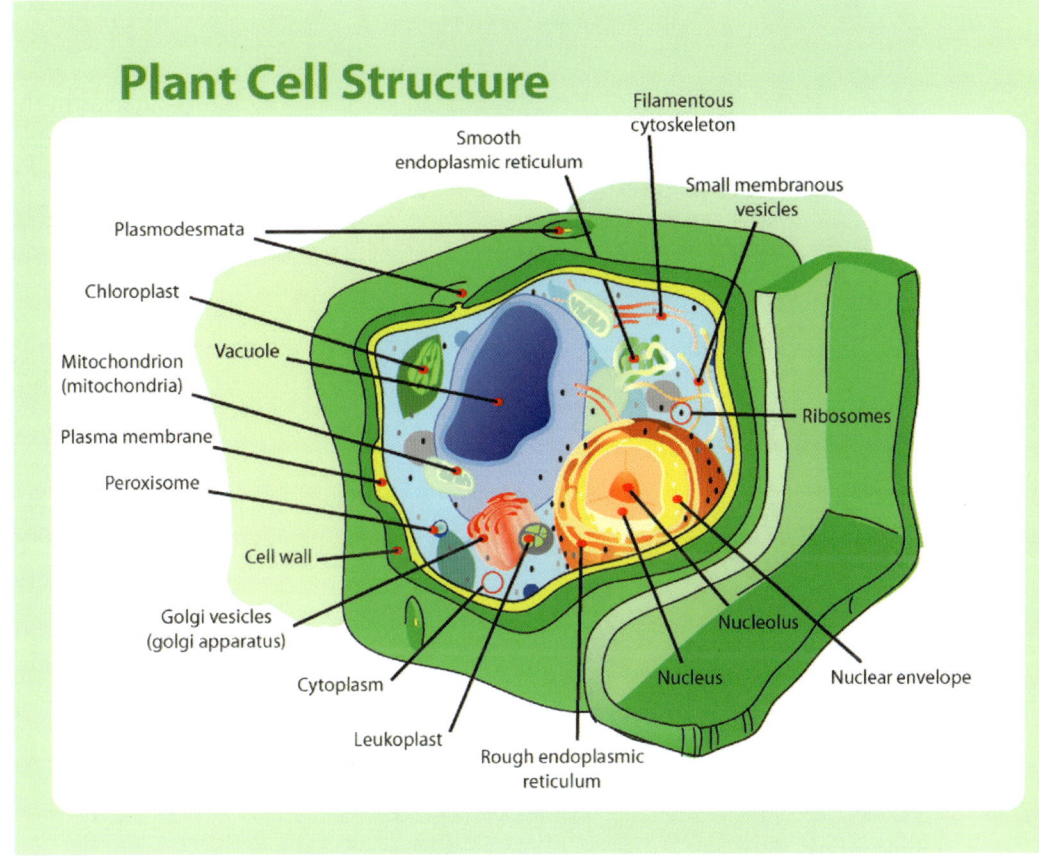

Plant cell structure showing the various structures present, including the cell wall and chloroplasts

Chloroplasts

Chloroplasts are very unique organelles found in plant and algae that absorb light and use it in the presence of water and carbon dioxide to produce sugars for energy production. They capture light energy in the form of chemical energy in ATP (adenosine triphosphate), which is ultimately used in many different reactions in the cell. The chloroplast is surrounded by a double lipid-bilayer membrane and contain its own DNA (60-100 genes), which can be seen under the microscope as flat discs. Within the chloroplast are stacks of membranes called thylakoids that contain light-absorbing pigments (chlorophyll and carotenoids). The pigments absorb energy from light and ultimately transfer it to water to release oxygen and in the process produce sugars.

Eukaryotic cell (plasma) membranes differ between animal, plant, and fungal cells. Animal cells are surrounded by a double membrane that is flexible, allowing the cells to adopt a variety of shapes. For example, there are approximately 210 distinct cell types in the human body. Plant cells are surrounded by a primary cell wall containing cellulose, hemicellulose, and pectin. This cell wall is very rigid and provides tremendous strength to the plant cell. Fungi contain a cell wall composed of chitin, while bacterial cell walls are composed of peptidoglycans. Inside the cell wall these plant and fungal cells contain a cell membrane similar to animal cells.

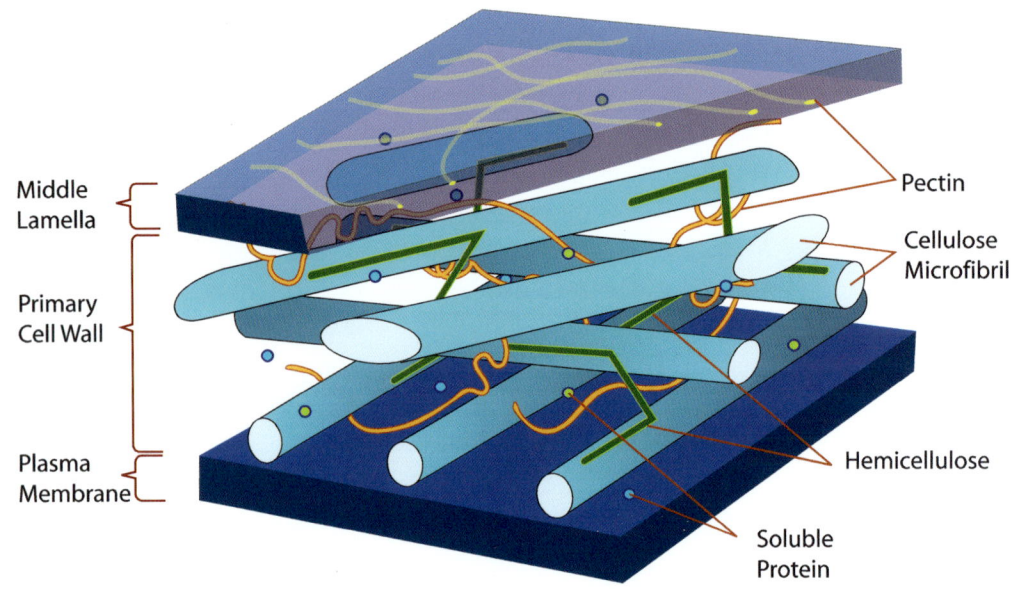

Plant cell wall structure showing cellulose, hemicelluloses, and pectin, along with other structural components

Eukaryotic cells, animals, yeast, plants, and prokaryotic cells (bacteria) are very important in biotechnology and have important roles in past and future developments. For example, they are used to make proteins, screen for new drugs, produce drugs, clean up the environment, or develop fuels.

Summary

Many differences exist between prokaryotic cells and eukaryotic cells, just as many differences exist between different types of eukaryotic cells (plant cells versus animal cells). These differences have allowed for all of these cell types to have distinct advantages in different environments and to evolve different specialized functions.

Concept Reinforcement

1. List several differences between eukaryotic and prokaryotic organisms and how they influence the organism.

2. What differences exist between mammalian animal cells and plant cells?

3. What components are present in the plant cell wall? What about the fungal cell wall? What are their functions?

4. What is the purpose of chloroplasts in plant cells?

Section 2.7 – Important Biomolecules in Biotechnology

Section Objective

- Discuss important biomolecules and how they are utilized in biotechnology

Biomolecules and Biotechnology

As discussed in unit 1, biotechnology can be described as a blended field of science and industry that includes the study of organisms and their molecular building blocks (**DNA, RNA, proteins, carbohydrates,** and **lipids**). These molecules might be used to benefit different areas, such as food production, farming, drug development, environmental cleanup, energy production, and almost any other field in which living things are a necessary part. The difference between basic scientific research into organisms and their molecular components and biotechnology is the applied nature of biotechnology, such that these organisms or biological components are utilized to make a product or process for human use. The key to biotechnology is these living organisms, including animals, plants, bacteria, yeast, or components of the organisms. This aspect is what gives biotechnology the "bio!" **Biochemistry** is the study of the chemical processes in living organisms and thus these biomolecules.

Let us consider the different types of biomolecules in a cell, and how they may be important in biotechnology.

DNA and RNA

DNA (deoxyribonucleic acid) and RNA (ribonucleic acid) are called nucleic acids and have very similar chemical makeup and structure. They are based on a sugar-phosphate backbone with nucleotide bases attached. The nucleotide bases can be thought of as decorating the sugar-phosphate backbone at regular intervals. The difference between DNA and RNA is the sugar involved: deoxyribose for DNA and ribose for RNA. DNA exists in a double-stranded structure, with the bases pairing with each other in the interior of the double strand because of its helical structure. Adenine (A) always pairs with thymine (T), while cytosine (C) always pairs with guanosine (G). RNA contains the base uracil instead of thymine and exists in a single-stranded form instead of double-stranded.

Biochemistry

The study of chemical processes in living cells and organisms; determines structure and function of all of the biomolecules present in different cell types and organisms.

DNA

The chemical constituent of the nucleus that makes up genes – the molecular basis of heredity. DNA is constructed of a double helix and has a sugar-phosphate backbone with purine and pyrimidine bases. The sequence of bases is what makes up a particular gene.

RNA

RNA is very similar to DNA, but in the cell it is usually single stranded and it contains a different sugar component than DNA. It also contains one different pyrimidine base than DNA. RNA acts as the template for protein production.

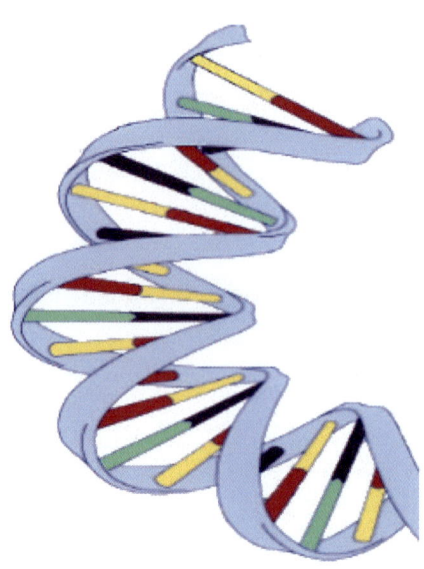

Illustration of the DNA double helical structure. The sugar-phosphate backbone is shown in blue and the nucleotide bases in red, yellow, green, and black.

Detailed chemical structure of RNA showing the sugar-phosphate backbone and the nucleotide bases A, G, C, or U attached. Note that RNA is single-stranded while DNA is double-stranded.

Gene

The section of DNA that codes for a protein. Genes are found in the genomic DNA.

Transcription

The process by which an RNA molecule is made from the genomic DNA template.

Translation

The process by which an RNA molecule is converted into a protein.

DNA contains all of the genetic information for a cell and is copied and passed from one cell to the next as cells replicate and divide. It contains the blueprint for all of the necessary information that a cell needs to live, grow, replicate, and die. This information is contained in separate **genes**, with each gene containing the information to produce a protein. While the proteins in a cell do all of the work, the DNA contains all of the information necessary to make them. The intermediary between the information stored in the DNA and the actual synthesis of a protein is the RNA. RNA is made in the nucleus from the DNA template (**transcription**), but then travels into the cytoplasm where it is used as a template for protein production (**translation**).

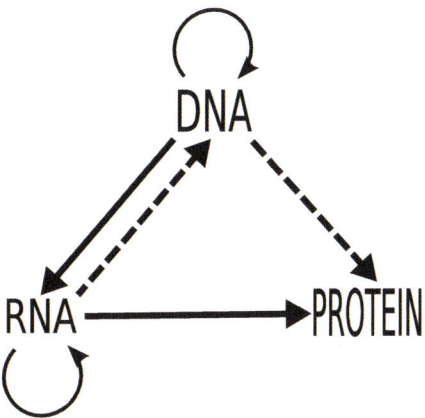

Central dogma of molecular biology: DNA codes for RNA which codes for protein

Because DNA contains the information for making a protein in the form of genes, the gene from one organism can be transferred to another, allowing the new organism to potentially make a new protein. The manipulation and transfer of DNA is the heart of modern biotechnology and involved in making proteins, industrial enzymes, **biotherapeutic** drugs, and **genetically modified organisms** (GMOs) to name just a few.

Proteins

Proteins are composed of strings of building blocks called **amino acids** and are formed during a process called translation from a messenger RNA (mRNA) template. The mRNA molecules contain the code needed to determine what order of amino acids is joined together to produce specific proteins. Proteins range in size from very small (a few tens of amino acids together such as some signaling proteins) to quite large with thousands of amino acids joined together. An example is titin, a protein containing 34,350 amino acids which is involved in muscle contraction in striated muscle tissue. This allows for the huge diversity of proteins within a single cell and allows them to fulfill numerous functions within that cell. Proteins provide the structural components of the cell or organism, both externally and internally, and are responsible for generating and using food and energy, making new molecules of DNA, RNA, sugars, and lipids, cell movement, and generally all functions within a cell or organism. Their importance can not be emphasized enough!

Biotherapeutic

A biomolecule that is used to treat a disease. These are usually proteins produced through biotechnology.

Genetically Modified Organism (GMO)

An organism that has had its DNA modified in some way. Genetic modification is often done to introduce an advantageous trait to plants.

Protein

Small to large molecules composed of amino acids. Proteins perform numerous essential functions in the cell and are very important in biotechnology.

Amino Acid

The basic building block of proteins. Amino acids are linked together during protein synthesis.

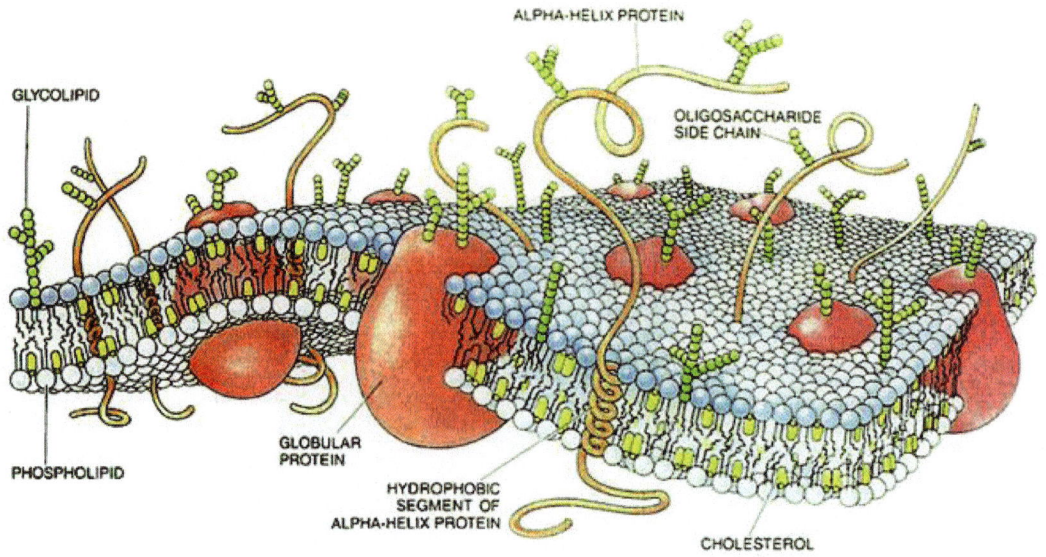

Eukaryotic cell membrane showing various components, including various proteins

As you might imagine, biotechnology finds many uses for proteins. They can be enzymes used in various industrial processes or in basic science research (i.e. tools of molecular biology). They can be hormones, growth factors, or antibodies that are used as biotherapeutic drugs. They can be used to coat surfaces for other biomolecules or cells to attach to or as targets for drug discovery, drug screening, or toxicity screening. They may be used as additives to foods to increase their nutritional value or to aid in digestion. In addition, they are used in the production of bioenergy and biofuels. The uses and potential uses for proteins in biotechnology are endless.

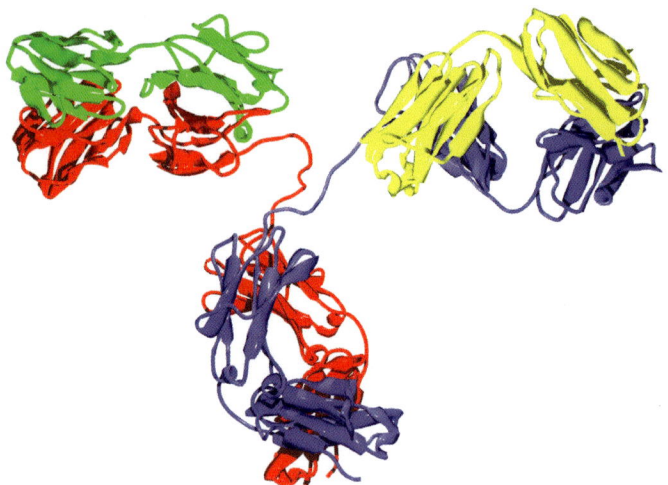

Molecular model of an antibody protein, which might be used as a biotherapeutic

Carbohydrates

Carbohydrates are the most abundant biomolecules in the cell and carry out numerous roles such as the storage and transport of energy, as well as providing structural components to the cell. They are composed of simple organic molecules ($C_nH_{2n}O_n$) with the basic building block being a monosaccharide, such as glucose, galactose, or fructose. Sugars are carbohydrates, but not all carbohydrates are sugars. These monosaccharides can be linked together to form polysaccharides in almost unlimited ways. Sucrose, or table sugar, consists of a glucose molecule and a fructose molecule linked together, while lactose consists of a glucose molecule linked to a galactose molecule (i.e. they are disaccharides). Two of the most common polysaccharides are the **cellulose** found in plant cell walls and **glycogen**, which is the form of energy storage in animals.

Table sugar (sucrose), which is a disaccharide of glucose and fructose

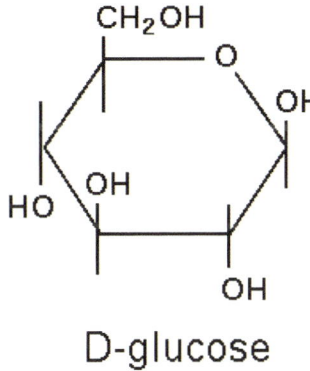

D-glucose

Chemical structure of D-glucose

Carbohydrates are important in biotechnology because many proteins used in biotechnology, particularly as **biotherapeutics**, are proteins that contain carbohydrates "decorating" them in various positions. These carbohydrate modifications are often essential to the function of the protein as well as help determine its **immunogenicity**. Because different types of eukaryotic organisms put different arrangements of sugar molecules on some of their proteins, the source of modified proteins is very important. Most biotherapeutic proteins that contain carbohydrates that are to be used in humans must be produced or manufactured in a human cell so they are compatible with each other.

Carbohydrate

A simple organic molecule composed of carbon, hydrogen, and oxygen.

Cellulose

A polysaccharide, composed of linked glucose molecules that is an important structural component of plant cell walls.

Glycogen

A polysaccharide composed of linked glucose molecules; it is the primary form of energy storage in animal cells.

Lipids

Lipids are a diverse group of molecules that refer to water insoluble or nonpolar compounds of biological origin and include such things as fats, fatty acids, waxes, phospholipids, sphingolipids, glycolipids, and various retinoids and steroids. Some lipids are linear in nature, while others contain ring structures. Some are flexible in their chemical structures, while others are quite rigid. They have diverse biological functions within a cell, but are the primary component of cell and organelle membranes. Some lipids have both polar and nonpolar characteristics, which makes them amphliphilic (having both **hydrophobic** and **hydrophilic** portions). Hydrophobic molecules do not like to be in an environment with water, while hydrophilic molecules do. They are also involved in cell signaling as second messengers in signal transduction pathways that control numerous important and critical cell functions.

Lipids also have a role in biotechnology. They are involved in the formulations of various drugs to make them available to the cells where they are needed, and may be key to various diseases and how they are treated. Lipids also play important roles in nutrition and nutritional biotechnology, and are critical to many industrial processes. Scientists are working on genetically engineering bacteria or other organisms to produce specific lipids for use by humans.

A free fatty acid

Cholesterol

A triglyceride

A phospholipid

Structure of four common lipids found in cells (fatty acid, cholesterol, triglyceride, and phospholipid)

Summary

There are many different types of biomolecules in each cell and organism, including DNA, RNA, proteins, carbohydrates, and lipids. Because of the differences in their chemical structures and functions, biotechnology takes advantage of these differences to produce unique and useful products.

Concept Reinforcement

1. What types of biomolecules are important in biotechnology and how are they different from each other?

2. Give an example of how proteins are used in biotechnology.

3. What role do carbohydrates play in biotechnology?

Section 2.8 – Chemistry and Replication of DNA

Section Objectives

- Describe the chemical structure of DNA and how it is organized in a cell

- Explain the process of DNA replication

Chemical Structure of DNA

Deoxyribonucleic acid (DNA) is the genetic material of all living organisms and even some **viruses**. Its main purpose is to provide the cell or organism the directions for constructing other cellular components, such as RNA and proteins, in a stable form that can be stored long term and copied for future daughter cells. The genetic information in DNA is organized into **genes**, which are further organized into very large DNA structures called **chromosomes**. The DNA is present in the nucleus of **eukaryotic** cells and in the cytoplasm of **prokaryotic** bacteria and **archaea** cells. In eukaryotic cells, the DNA is organized into a protein:DNA complex called chromatin, while during cell division the DNA condenses into actual chromosomes. There are two types of chromatin: 1) **euchromatin** is less compact and contains the frequently expressed genes (those copied into RNA), and 2) **heterochromatin**, which contains more tightly packed DNA that is not expressed as RNA.

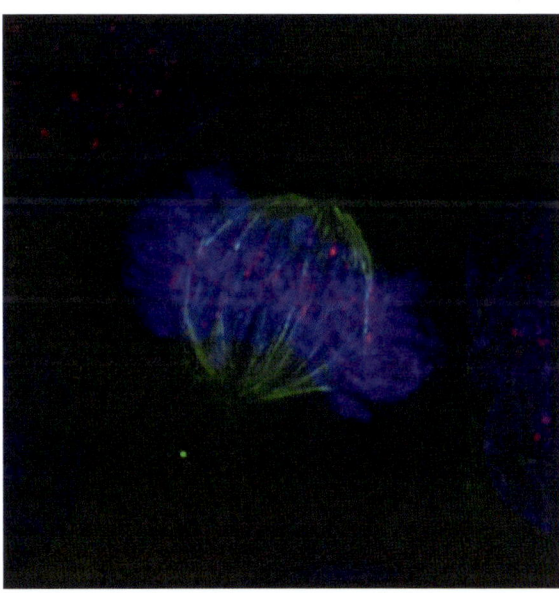

Cell in metaphase in mitosis (cell division) showing the copied and condensed chromosomes lined up to prepare for separation

---

Archaea

Single-celled organisms similar to bacteria, but they evolved separately and are actually more similar to eukaryotic cells.

Chromosome

A large DNA:protein complex.

Euchromatin

A less compact DNA in the nucleus of eukaryotic cells; where active gene expression occurs.

Eukaryotic

A cell having DNA contained within a separate organelle (the nucleus).

Gene

The smallest functional unit of DNA. Each gene contains the information for one RNA and thus one protein.

Heterochromatin

A more compact region of DNA in the nucleus. This is the region where active gene expression is not occurring.

Prokaryotic

Cells that do not contain a nucleus, so DNA is present in the cytoplasm.

Virus

A submicroscopic, infectious agent that infects host cells but cannot divide or reproduce outside a living cell.

Chemically, DNA is composed of a long string of units or building blocks called nucleotides, attached to a sugar and phosphate backbone joined by **ester bonds**. Attached to each 2-deoxyribose sugar molecule is one of the four nucleotide bases, termed adenine (A), guanine (G), cytosine (C), or thymine (T). In nature, DNA exists as two strands wound around each other to form a double helix or right-handed spiral. There are two grooves twisting around the surface of the helix; one groove (major) is wider than the other grove (minor groove).

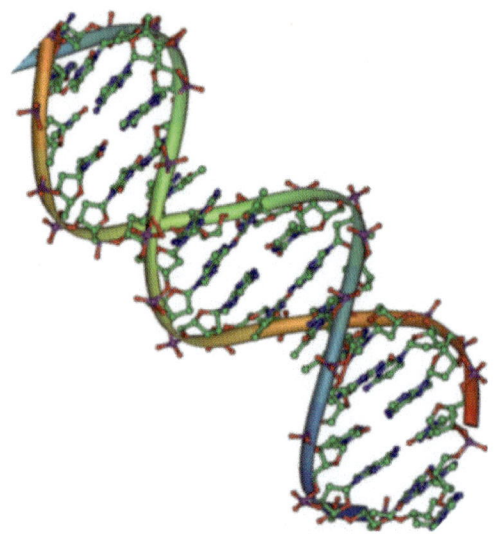

DNA double helix showing the nucleotide bases inside the helix with the sugar-phosphate backbone on the outside. Note the major (larger) and minor (smaller) grooves in the helix structure.

The two strands of the helix are held together by specific base pairing (hydrogen bonding) of the nucleotide bases: adenine always pairs with thymine and cytosine always pairs with guanine. This pairing is not permanent, so the two strands of the helix can separate when needed, such as when the DNA is being copied. It is like unzipping the two parts of a zipper and can be performed by mechanical force, heat, or specific proteins in the cell. Each DNA strand has directionality and when they pair, they do so such that their directions are opposite of each other (called **antiparallel**). In addition, when the nucleotides pair together to form a base pair such that adenine always bonds with thymine and guanine always bonds with cytosine, this binding is described as **complementary** and plays an important role when DNA copies itself (replication). The two types of base pairs form different numbers of hydrogen bonds: AT forming two bonds and GC forming three bonds. Thus the interaction between a GC pair is stronger than the AT pair.

The information in the DNA is stored in the specific sequence of base pairs in the different regions of DNA. Each gene has a unique sequence (like a barcode) as compared to other genes. Each gene in turn contains the information to make one RNA molecule, which in turn contains the information to encode for one protein. The process of DNA being transcribed into RNA, which in turn is translated into a protein, is called the central dogma of molecular biology (see diagram in section 2.7). Interestingly, of all of the DNA in a human cell, only about 1.5% actually contains information required to make proteins. The other DNA consists of repetitive regions (about 50%) and non-repetitive

128

sequences. The exact function of these non-coding regions is not fully understood, but they may play roles in gene regulation (when and where genes are expressed into RNA in the organism).

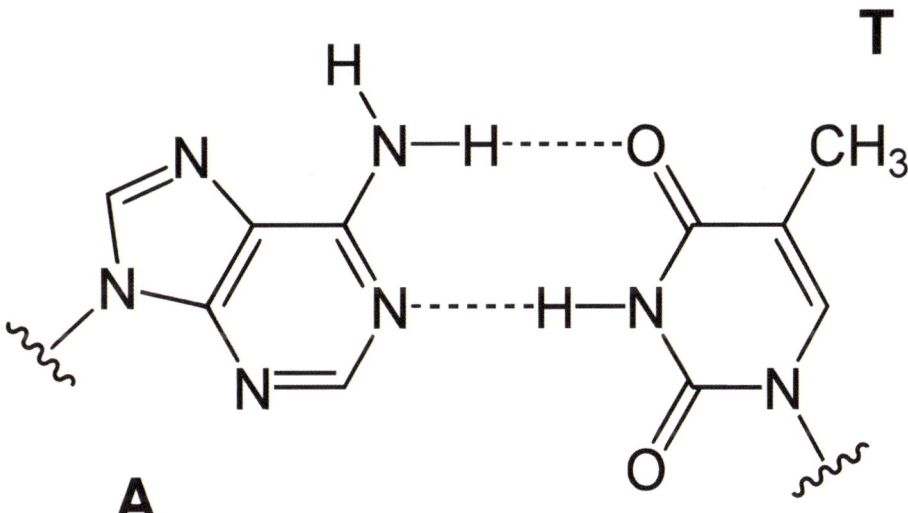

Guanine:cytosine base pair in DNA. Notice the three hydrogen bonds
that hold C and G together in the helix.

Adenine:thymine base pair in DNA. Notice the two hydrogen bonds
that hold A and T together in the helix.

DNA Replication

In order for a cell to grow and divide into daughter cells, the DNA must first replicate or copy itself. This process is called DNA replication and requires a number of protein helpers. It is carried out by living organisms and is the fundamental basis for inheritance. As each DNA strand holds the same genetic information, both strands can serve as templates for reproduction of the other strand. This is due to the complementary base pairing between AT and GC. The resulting double-stranded molecules are identical to the parent double-stranded molecule and proofreading mechanisms exist to ensure no mistakes are made in the process. If an error is found, other functions correct it.

DNA replication begins at special regions in the DNA called **origins**, where the two DNA strands separate. A short piece of RNA binds to the DNA strand and an enzyme called DNA polymerase extends from the piece of RNA making a new DNA strand. The unwinding of DNA and synthesis of new strands for both template strands form a **replication fork**. The resulting fork has two prongs, each made up of a single strand of DNA. One daughter strand of DNA is synthesized in a continuous manner, and is called the leading strand. The other strand of DNA is copied in a discontinuous manner, and is called the lagging strand. In addition to DNA polymerase, a number of other enzymes are involved in the process, including DNA binding protein (helps keep strands separated), nucleases (break down unwanted DNA), ligases (seal newly made DNA strands together in the lagging strand), topoisomerases (help untwist DNA as it is replicating), and helicases (help unzip DNA as it is replicating). Within a eukaryotic cell, DNA replication is regulated within the cell cycle. As the cell grows and progresses through the various stages of the cell cycle, it leads into DNA synthesis and ultimately cell division into two daughter cells.

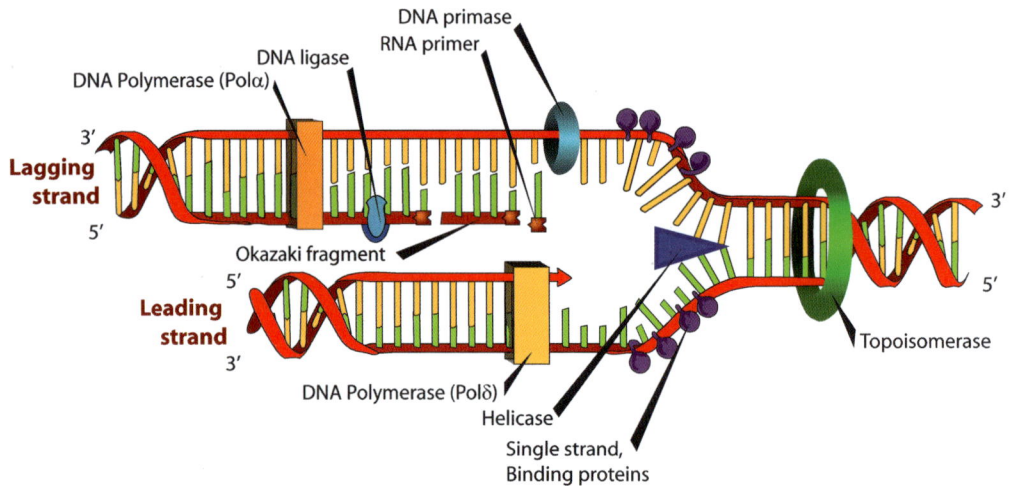

DNA replication showing the enzymes involved for both
the leading and lagging strand replication

- Pericentriolar material
- Mother centriole
- Daughter centriole
- Replicated DNA
- Chromosome

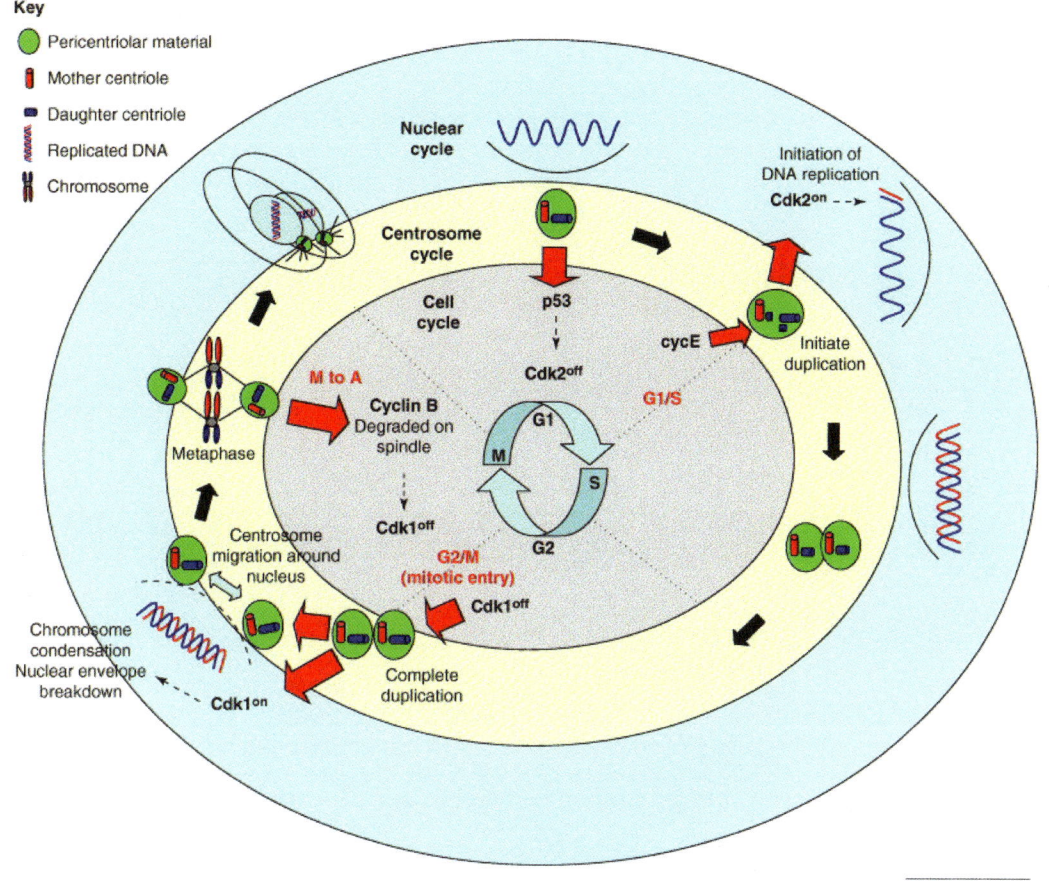

TRENDS in Cell Biology

Cell cycle, with the G1, S, G2, and M phases shown.
DNA replication occurs in the S (synthesis) phase.

Cell division in eukaryotes involves separating the duplicated chromosomes by microtubules that become attached to the center of each chromosome. In **mitosis**, one cell divides into two identical daughter cells, each with a complete copy of the genetic information (DNA). The cell cycle is divided up into four distinct phases: G1, S, G2, and M. In eukaryotic cells, each cell has two copies of each chromosome and is called **diploid**. One chromosome comes from the mother organism and one chromosome comes from the father organism during fertilization of an egg with a sperm cell. In **meiosis**, one cell divides but does not replicate its DNA first, so the daughter cells have only one copy of each chromosome and are called **haploid**. This process is how eggs and sperm are formed and is required for sexual reproduction.

Telomeres

At the ends of chromosomes are specialized regions of DNA called telomeres. Their main function is to allow the cell to completely replicate the ends of the chromosomes, so no genetic information is lost. These specialized regions also protect the end of the chromosomes from degradation (break down). In human cells, telomeres are usually lengths of single stranded DNA containing several thousand repeats of the sequence TTAGGG. They are added to the ends of the chromosomes by an enzyme called telomerase.

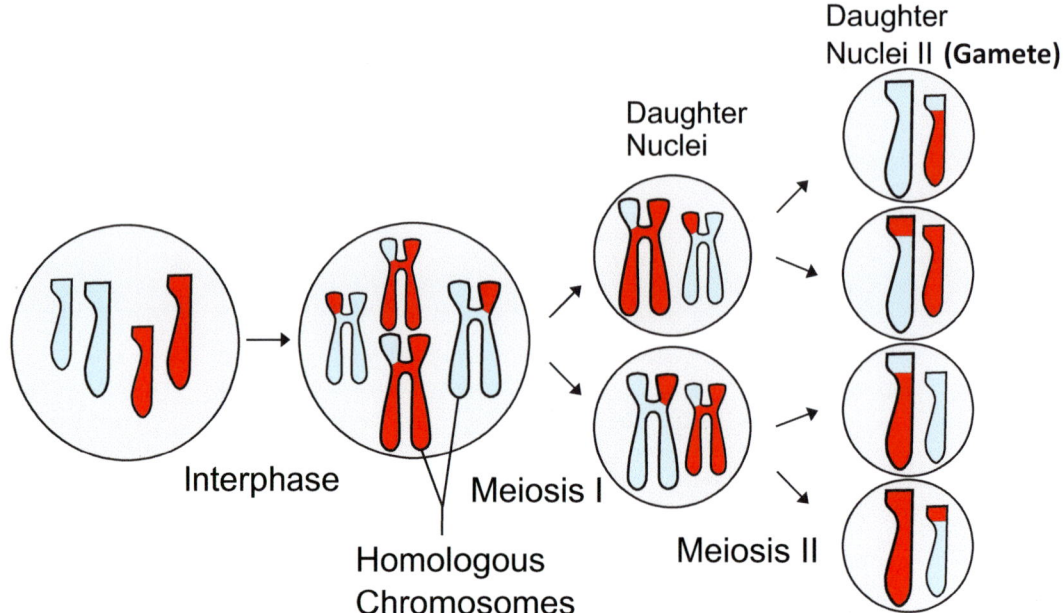

Meiosis showing the separation of diploid chromosomes
into gametes with only haploid chromosomes

Term	Definition
Diploid	A cell having two copies of each chromosome.
Haploid	A cell having only one copy of each chromosome.
Mitosis	The complete duplication of DNA and separation into two daughter cells.
Meiosis	The process by which duplicate copies of chromosomes segregate into separate daughter haploid cells (gametes).
Telomerase	The enzyme that adds special DNA sequence to the ends of chromosomes following replication.
Telomeres	The ends of chromosomes composed of special repeated sequences; act to protect the ends of DNA from degradation and to allow for complete copying during replication.

While mutations can arise during DNA replication, other mechanisms also exist to increase genetic variation within a population of organisms. One such event is genetic recombination. In this process, two DNA helices break, swap sections, and then rejoin. Recombination allows chromosomes to exchange genetic information and produce new combinations of genes, which increases the efficiency of natural selection, contributing to evolution of a species.

Summary

Thus, DNA replication is critical to cell growth and cell reproduction and the chemical structure of DNA makes its replication mechanism possible.

Concept Reinforcement

1. How is DNA packaged in a cell?

2. Describe the chemical structure of DNA.

3. How is DNA replicated prior to mitosis?

Section 2.9 – DNA Purification and Analysis

Section Objectives

- Describe the DNA purification process

- Discuss various methods that are used to analyze DNA

- Explain what role DNA plays in biotechnology

DNA Purification

DNA is often manipulated in biotechnology for use in the analysis of specific genes, recombinant (artificial or laboratory) protein production, the production of DNA for **gene therapy**, or for making **genetically modified organisms** (GMOs). The DNA of interest could be genomic DNA from an organism, plasmid DNA from bacteria, or viral DNA from an infectious and pathogenic virus. The first step in all of these processes is the purification of the target DNA for subsequent analysis and manipulation. This involves first lysis (rupture) of the target cell or organism to remove the DNA from the nucleus if the cell is eukaryotic or from the cytoplasm if prokaryotic. This step is usually accomplished using a solution that contains high amounts of **guanidine** salt, a **detergent**, or both. The guanidine salt acts as a **chaotropic** agent to strip the water molecules off of the DNA to allow it to bind to other substances that can be used in purification. It also acts to disrupt the three-dimensional shape of DNA, RNA, and proteins, simplifying the purification process. The detergent acts to solubilize (melt) the lipid-based cell membrane of the cell and nucleus to release the cellular contents for isolation.

Purified DNA floating in ethanol (note the whitish strands
of DNA near the bottom of the flask)

<aside>

Chaotropic

Disrupting the structure of water, macromolecules, or a living system to allow activities that would normally be inhibited by the structure.

Detergent

A substance that can solubilize membranes or lipids.

Gene therapy

The insertion of a gene into a person's DNA to treat or cure a disease, usually a protein that is either absent or abnormal through hereditary defects.

Genetically modified organisms (GMO)

A living thing that has had its DNA altered, usually with the addition of genetic material from another organism that confers a desirable trait.

Guanidine

Strongly alkaline crystalline compound formed by the oxidation of the nucleotide base guanine. It has strong chaotropic properties and can denature proteins.

</aside>

Once the DNA is released from the cell, often the **denatured** proteins are removed from the lysate (cells that have been broken open) by **centrifugation**. This step also removes other larger cellular debris, leaving the target DNA in the liquid phase. The DNA may be captured on a **silica**-based solid support, such as a membrane or magnetic bead, and following a number of wash steps to remove contaminants and impurities, it can then be released from the silica in a pure form in water. The DNA may also be precipitated out of the lysate using an alcohol such as ethanol. DNA is not soluble (does not dissolve) in alcohol, so it becomes a solid (precipitates). DNA is soluble in aqueous (water-based) solutions, and thus exists in the solution and not as a separate solid. Purification of plasmid DNA from bacteria is similar, but since plasmid DNA is so much smaller than genomic DNA, the large chromosomal DNA pellets in the debris fraction and the smaller plasmid DNA is present in the liquid and ready for purification using silica. Many biotechnology reagent and research tool companies develop and sell kits for the purification of various biomolecules, including DNA.

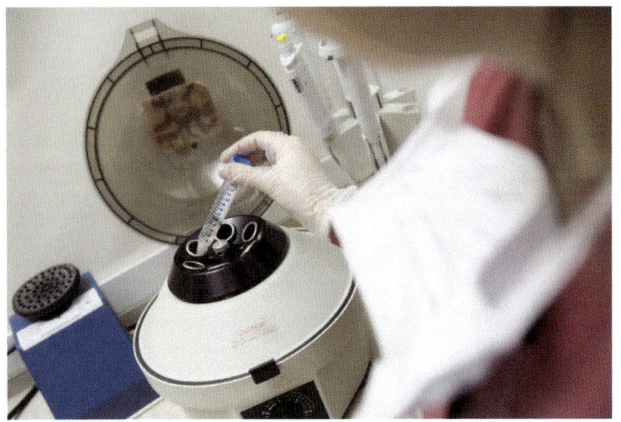

Microcentrifuge used for DNA purification

Average amount of DNA that can be purified from various organisms or cell types.

Sample Type	Amount of Sample (# of cells or milligrams of tissue)	Average Yield of Genomic DNA
Human cell	Per one cell	~6 picograms
Human cell	1 million cells	~5-10 micrograms
Human blood	1 milliliter (~4-7x10⁶ cells)	~25-60 micrograms
Animal tissue	10 milligrams (0.01 gram)	~15-20 micrograms
Plant leaf tissue	10 milligrams (0.01 gram)	~2-3 micrograms
Bacteria	Per one cell	~0.017 picogram
Bacteria	10⁹ cells	~10-20 micrograms
Yeast	10⁹ cells	~25-35 micrograms

DNA Analysis

Once the DNA is purified from the target cell or organism, it then can be analyzed in a variety of ways. Minimally, scientists are interested in knowing how much DNA is present and its purity. They may also want to verify that the purified DNA is not degraded (broken down), but intact and of good quality. Many downstream (post-purification) applications require that the DNA be highly pure and at a certain concentration. Measuring the amount and purity of DNA can be accomplished by assessing the **absorbance** at 260 **nanometers**. Many biological molecules absorb light at specific wavelengths, and DNA and RNA absorb light in the ultraviolet range at a specific wavelength of 260 nanometers. In contrast, proteins absorb ultraviolet light at 280 nanometers. By measuring the amount of light absorbed at these wavelengths, using a special machine called a **spectrophotometer**, the concentration of DNA can be calculated using the **Beer Lambert Law**: the concentration of DNA in a solution is directly proportional to the absorbance at 260 nm. The concentration of DNA is calculated using the equation (Absorbance at 260nm) x (50ug/ml) = concentration of purified DNA in micrograms/milliliter. If a dilution is made of the purified DNA prior to recording the absorbance measurement, then this dilution factor will have to be taken into account when calculating the final DNA concentration.

The purity of the DNA can be estimated by calculating the ratio of the absorbance at 260nm (DNA) divided by the absorbance at 280nm (protein). This ratio gives an indication of the amount of protein contamination in the purified DNA sample. A ratio of > 1.7 to < 2.1 is considered of high purity and acceptable for later applications.

In addition to absorbance as a method for quantifying DNA, other methods exist that utilize the ability of fluorescent molecules to bind to DNA and give a quantitative signal that can be measured with a fluorescent reading machine (fluorometer). These methods are very sensitive and quantitative, but do not provide any information about the purity of the DNA sample.

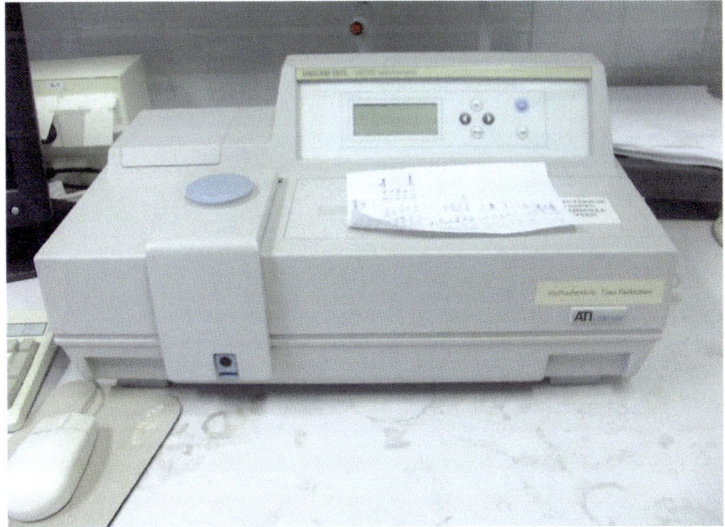

Typical spectrophotometer used in a laboratory to measure the amount of light absorbed by the DNA in solution

---

Absorbance

The ability of various substances to absorb light at particular wavelengths.

Beer Lambert Law

The mathematical calculation that allows for the calculation of DNA concentration from absorbance at 260nm.

Nanometers

A unit of measure for length. One nanometer is equal to 10^{-9} of a meter.

Spectrophotometer

A device for measuring light intensity as a function of the wavelength of light; this machine can measure the amount of light absorbed by a substance when placed in the path of the light beam.

USPTO

United States Patent and Trademark Office.

Once the DNA has been quantified and determined to be of high quality, it may then be subjected to other manipulations. To determine the size of the purified DNA molecules, agarose gel electrophoresis is often employed. This technique involves subjecting DNA to an electric field in a solid support composed of an agarose gel or matrix. Agarose is a molecule derived from seaweed that when dissolved in water will solidify into a gel-like solid, similar to Jell-O. The DNA can then be loaded into the agarose solid support. When subjected to an electric current it will migrate into the matrix (mesh) according to its size, as DNA is negatively charged and will migrate toward the positively charged electrode (cathode). Smaller pieces of DNA will more easily migrate through the gel and thus travel farther down the gel to the cathode. Larger pieces of DNA will be retarded in the gel and not be able to travel as far down the gel toward the cathode. Standard pieces of DNA of known size are also analyzed concurrently with the purified DNA samples to allow for size determination. These standard pieces of DNA are called molecular weight markers or ladders and their sizes are expressed in the number of base pairs (bp) in the piece of DNA. High quality genomic DNA, which is intact and not degraded from eukaryotic cells, should be at least 10-30 kilobase pairs (kbp) in length. The correct size of a DNA fragment can be indicative of the successful purification of the DNA of interest.

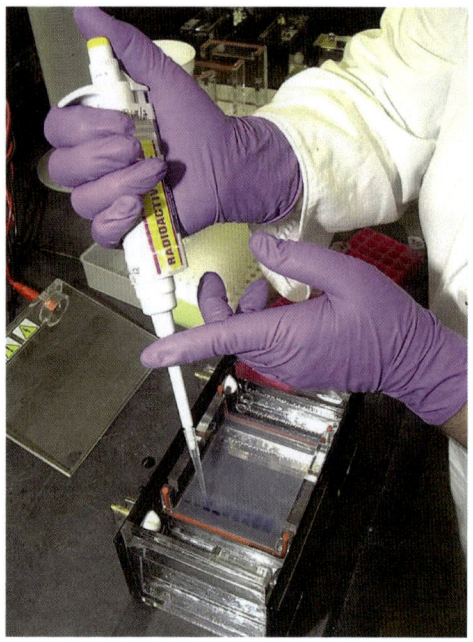

Scientist loading a DNA sample into an agarose gel preparing for electrophoresis analysis

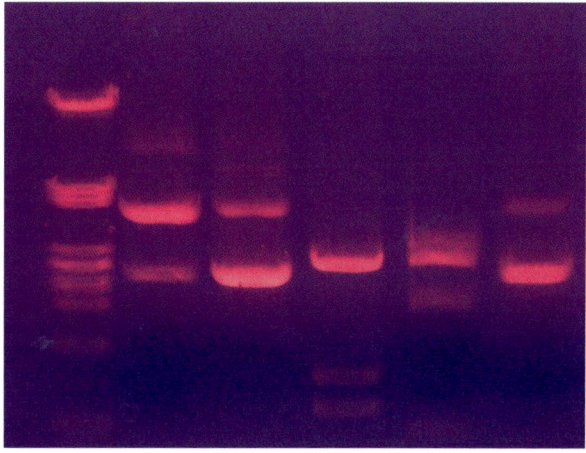

DNA bands present in an agarose gel following electrophoresis.
Note the bands in the molecular weight marker in lane number 1,
with other DNA samples present in lanes 2 – 6.

DNA in Biotechnology

The purification and manipulation or analysis of DNA from various organisms is the framework and basis for modern molecular biology and biotechnology. The target DNA may contain the information to express a protein that is useful for industrial processes or has therapeutic value (like a drug). The DNA may be used to identify individuals in the case of forensic analysis or in the investigation of evolutionary **phylogeny** and population migration.

DNA is even being used in **nanotechnology**, which takes advantage of the specific interactions between different molecules of DNA at the molecular level to serve as a structural material.

The unique structure and physical characteristics of DNA make it an ideal molecule for storing all of the hereditary information, as well as for being used as a tool in biotechnology.

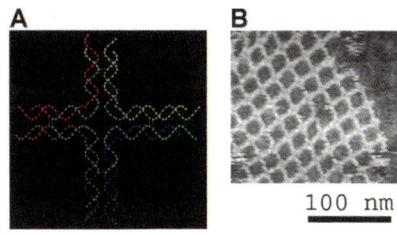

Self-assembled DNA nanostructures such as these nanogrids

Nanotechnology

The broad field of science that deals with the control of matter on the molecular or atomic scale, usually 100nm or smaller, and the manufacture of devices with dimensions that are in that range.

Phylogeny

The study of the evolutionary relatedness between organisms, usually based on DNA or protein sequences.

Summary

DNA is purified primarily by the ability of silica to bind to the negative charge on the DNA. In this way the DNA can be separated from all of the other cellular constituents. The purified DNA can be analyzed for quantity and quality (integrity) using several different methods including absorbance and agarose gel electrophoresis.

Concept Reinforcement

1. Describe the process of DNA purification using a silica-membrane based method.

2. What characteristic of DNA allows it to be quantitated using a spectrophotometer and what wavelength is involved?

3. What characteristic of DNA does agarose gel electrophoresis measure and how does it do so?

Section 2.10 – Chemistry and Transcription of RNA

Section Objectives

- Describe the chemical structure of RNA and list the different types of RNA

- Explain the process of transcription (the link between DNA and protein)

Chemical Structure of RNA

Ribonucleic acid (RNA) is the link between the information encoded in the DNA to the expression of proteins in a cell. RNA is chemically similar to DNA in many respects. For example, it includes a sugar-phosphate backbone, with nucleotide bases attached to the sugar molecules. In the case of RNA, the sugar component is ribose instead of deoxyribose and the four nucleotide building blocks include adenine (A), guanine (G), cytosine (C), and uracil (U) instead of thymine (T). Unlike DNA, RNA exists in a single strand and not in a double-stranded helix. However, because of the complementarity between bases and the ability of guanine to hydrogen bond with cytosine and adenine to hydrogen bond with uracil, it can bind to itself to form various secondary structures such as hairpins, loops, bulges, and internal loops. The ability of RNA to adopt such novel structures by binding to itself has allowed some RNAs to accomplish chemical changes, similar to an enzyme (called **ribozymes**).

> **Ribozyme**
>
> An RNA molecule that can actually accomplish a simple chemical reaction, similar to an enzyme.

Chemical structure of RNA showing the sugar-phosphate backbone and attached nucleotide bases A, G, C, and U

Because of the different sugar molecule on RNA versus DNA, RNA is forced to adopt a different geometry than DNA. This difference in sugars also makes RNA much less stable than DNA, causing it to be more prone to break-down (hydrolysis), particularly in **alkaline** conditions.

> **Alkaline**
>
> Having pH greater than 7.0 (basic).

In addition to the unique nucleotide base uracil in RNA, numerous modified bases and sugars can be found in nature in mature RNA molecules. Many of these modified bases occur in transfer RNAs, which will be discussed later in this section and are crucial to protein expression (translation).

Types of RNA in the Cell

There are three main types of RNA in the cell (messenger RNA, ribosomal RNA, and transfer RNA), each with different functions. However, many other types of RNA exist that may be specific to eukaryotes or prokaryotes or specific to a type of organism (see table).

Messenger RNA

The RNA responsible for serving as the template in protein synthesis.

Messenger RNA (mRNA) constitutes about 1-3% of the total RNA within a cell and serves as the template for translation (protein synthesis). It contains the sequence of bases that designate which amino acids are to be incorporated during translation into the growing protein chain. Messenger RNAs also contain features that are recognized by **ribosomes** and other translation factors that aid in the efficiency of translation. mRNAs can be of various sizes, since proteins come in different sizes, but range between a few hundred bases to several thousand bases.

Ribosome

Large RNA:protein complex responsible for the translation process.

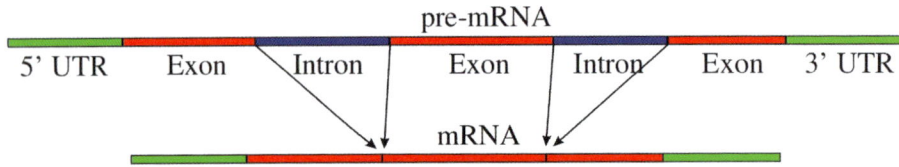

| 5' UTR | Exon | Intron | Exon | Intron | Exon | 3' UTR |

pre-mRNA

mRNA

Diagram illustrating messenger RNA (mRNA) before and after processing to the mature form

Ribosomal RNA

Integral RNA component in the ribosome that is crucial for translation.

Ribosomal RNA (rRNA) constitutes the vast majority (~85%) of RNA within a cell and is an integral component of the ribosome, an organelle that performs the process of translation. **Transfer RNAs** (tRNAs) are the link between the sequence of bases in the RNA and the corresponding amino acid that it codes for. Transfer RNAs constitute ~10-15% of the total RNA within a cell and contain an anticodon loop that is complimentary to the codon sequence in the mRNA. Transfer RNAs are very small, containing only ~80 nucleotides, but carry the appropriate RNA to the ribosome for translation. There are four rRNA molecules in eukaryotic cells and three in prokaryotic cells. The small ribosomal subunit containing 1 rRNA molecule and the large ribosomal subunit containing 3 rRNA molecule in prokaryotic cells and 4 rRNA molecules in eukaryotic cells. In addition, eukaryotic cells have two extra rRNAs in the mitochondria. All of the various rRNA molecules are differentiated by their size, which ranges from 120 nucleotides to 5,000 nucleotides in length.

Transfer RNAs

Link between mRNA and protein translation, as they carry and match the correct amino acid to the corresponding mRNA codon.

142

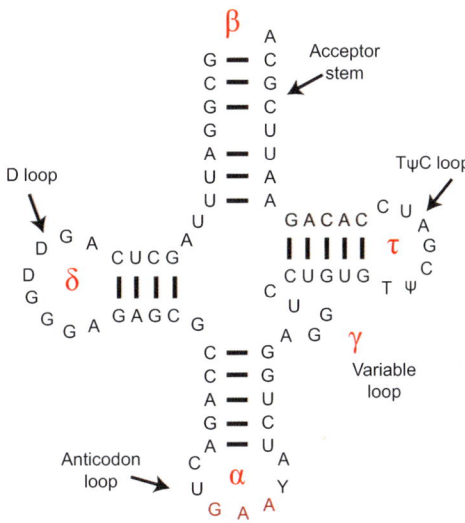

Typical transfer RNA molecule. Note the anticodon loop at the bottom of the T-like structure. The acceptor stem is where the amino acid is attached.

Other types of RNA found in eukaryotic cells allow for the regulation of gene expression. That is, when, where, and how much of a particular mRNA is made, and thus, usually how much of the protein is made. The regulation of gene expression is critical to the development and life of any cell or organism, and involves many complex mechanisms, some of which involve RNA components. These include small interfering RNAs (siRNAs), microRNAs, and transacting siRNAs. These small RNAs either inhibit translation of the mRNA or target the mRNA for premature degradation (break down). Through either mechanism, they effectively cause a decrease in the amount of protein made from a particular mRNA.

Type of RNA	Function	Distribution
Messenger RNA (mRNA)	Codes for proteins	All organisms
Ribosomal RNA (rRNA)	Translation (protein synthesis)	All organisms
Transfer RNA (tRNA)	Translation (carry amino acids)	All organisms
Transfer-messenger RNA (tmRNA)	Rescuing stalled ribosomes during translation	Bacteria
Antisense RNA (asRNA)	Gene regulation	All organisms
Small interfering RNA (siRNA)	Gene regulation	Most eukaryotes
MicroRNA (miRNA)	Gene regulation	Most eukaryotes
Small nuclear RNA (snRNA)	Various functions	Eukaryotes and Archaea
Small nucleolar RNA (snoRNA)	Nucleotide modifications of RNAs	Eukaryotes and Archaea
Telomerase RNA	Telomere synthesis	Most eukaryotes
Signal recognition particle RNA (srpRNA)	Protein export	All organisms
Trans-acting siRNA (tasiRNA)	Gene regulation	Plants

Small Nucleolar RNA (snoRNA)

Small nucleolar RNAs are a class of small RNA molecules that act as guides for chemical modifications of ribosomal RNAs (rRNAs) or small nuclear RNAs (snRNAs). They are commonly referred to as guide RNAs, so that the correct nucleotides in the ribosomal RNA are modified. Each snoRNA molecule can act as a guide for only one or two individual modifications in the target RNA. A number of different proteins are involved in these modifications and they interact with both the snoRNA and the target RNA. These chemical modifications of the target RNAs are crucial for their function, and thus snoRNAs are very important in the cell.

Transcription

All RNAs within a cell are generated through a process called transcription from blueprints found in the DNA. Severo Ochoa won the Nobel Prize in Medicine in 1959 for the discovery of how RNA is synthesized, and Roger Kornberg won the Nobel Prize in Chemistry in 2006 for his studies of the molecular basis of eukaryotic transcription. Specific regions in the DNA called **promoters** are recognized by enzymes called **RNA Polymerases** that bind to the DNA and make a complimentary copy from one of the DNA strands. Unlike the enzymes that copy DNA during DNA replication, RNA polymerases do not need a primer molecule to start transcription. Prokaryotic cells have one RNA polymerase enzyme, while eukaryotic cells contain three RNA polymerase enzymes: 1) RNA Polymerase I, which transcribes rRNAs, 2) RNA Polymerase II, which transcribes mRNAs, and 3) RNA Polymerase III, which transcribes tRNAs and other small RNAs.

The process of transcription involves three main steps: initiation, elongation, and termination. During initiation, RNA polymerase and other accessory proteins bind to the promoter region of the DNA and start transcription. The process in eukaryotes and archaea is much more complicated than that in prokaryotes, as many more protein cofactors are involved in this highly regulated process. One strand of DNA, the template or antisense strand, is then copied into an RNA copy during the elongation phase. The RNA polymerase uses base pairing complementarity to create the correct sequence of bases in the RNA copy (called the sense strand). During transcription, thymine is replaced with uracil and the other nucleotides contain the sugar ribose instead of deoxyribose. Nucleotides used in DNA synthesis are called deoxyribonucleotides (dNTPs), while those used in RNA synthesis are called ribonucleotides (rNTPs or NTPs). Termination occurs when the RNA polymerase recognizes certain features in the RNA, in conjunction with other protein factors in bacteria, and stop transcription. Many copies of an RNA can be made from a template DNA, since the RNA is not damaged in any way during transcription. As with DNA replication, there are mechanisms present during transcription that can correct mistakes that might be made by RNA polymerase.

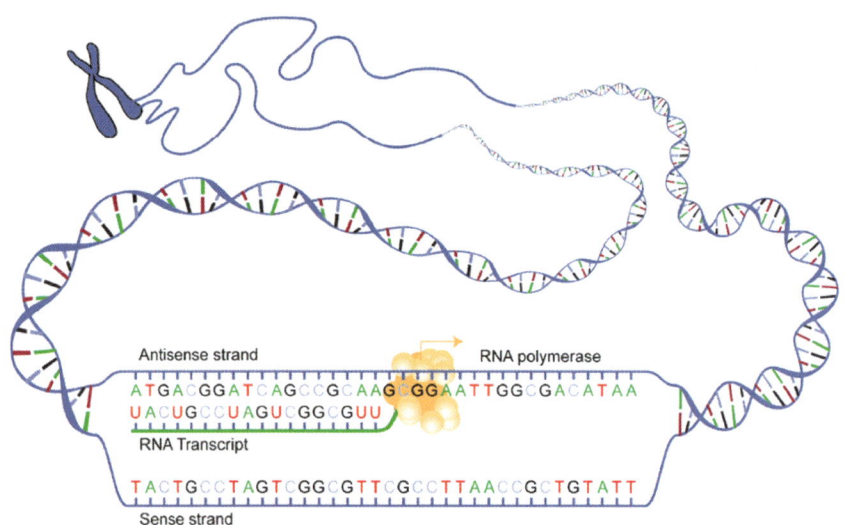

Transcription of the DNA template by RNA polymerase to produce an RNA strand.

Eukaryotic RNAs, once transcribed, then go through a process of splicing during which

144

some of the regions are removed (**introns**), leaving only the sequence that directly codes for a protein (**exons**). Splicing occurs in the nucleus of eukaryotic cells, and the processed mRNA is then exported to the cytoplasm, where translation takes place. Prokaryotic messenger RNAs do not contain introns, so are not processed prior to being translated into a protein. This means that in bacteria transcription and translation can actually occur simultaneously. Processed eukaryotic mRNAs are also further modified by the addition of special nucleotides to each end of the mRNA to form a mature mRNA molecule.

Intron

A nucleic acid sequence that is not present in the mature mRNA. Introns are removed prior to translation, during the process of splicing in the nucleus.

Exon

A nucleic acid sequence present in the mature mRNA. An exon is a specific part of the RNA sequence that codes for a protein.

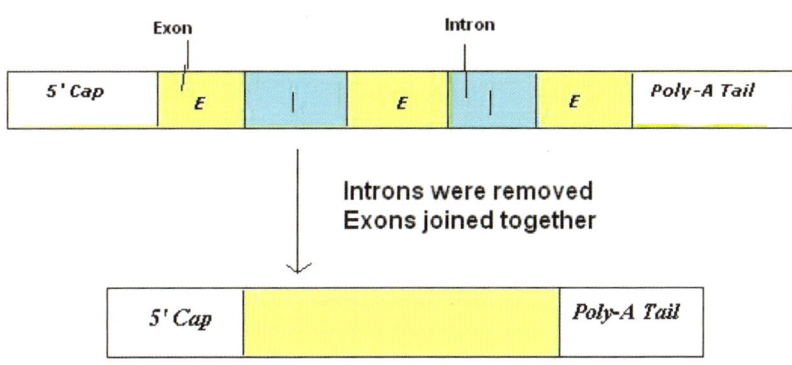

RNA splicing

RNA processing step called splicing, in which introns are removed, leaving the exon sequences in the RNA. The mature mRNA also contains a 5' cap and a polyA tail.

Some viruses have the ability to transcribe an RNA molecule into a DNA copy, called complimentary DNA (cDNA). This process is called reverse transcription and is crucial to the life cycle of retroviruses. An example of a retrovirus is **HIV**, which causes **AIDS**. Some of the drugs developed to treat AIDS are chemicals that inhibit the reverse transcriptase enzyme of HIV, since normal human cells do not contain this enzyme.

RNA transcription is a complicated and highly regulated process, since the amount of any particular protein can dramatically affect the health and response of a cell to its internal and external environments.

AIDS

Acquired immune deficiency syndrome, which is caused by HIV infection.

HIV

Human immunodeficiency virus. The cause of AIDS.

Summary

Ribonucleic acid (RNA) plays a very important role in the cell. It allows for new proteins to be made by carrying the information in the DNA out to the cytoplasm and also serves as the actual template for protein synthesis. RNA is made in the cell through a process called transcription, which involves an enzyme called RNA Polymerase.

Concept Reinforcement

1. How are the chemical structures of RNA and DNA similar? How are they different?

2. What are the three main types of RNA in a cell and what are their functions?

3. How is RNA made and what steps are involved in the process?

Section 2.11 – Purification and Analysis of RNA

Section Objectives

- Describe the basic methods for purifying RNA

- Describe various methods for analyzing RNA

- Explain what role RNA plays in biotechnology

RNA Purification

There are many different methods for the purification (isolation) of RNA. As you might expect due to their chemical similarities, these methods tend to be very similar to those used for DNA purification. In reivew, the first step is lysis of the target cell, organism, or tissue, again using high concentrations of **guanidine** salt and other the presence of a **detergent**. The high amount of guanidine salt acts not only to lyse the cells, but quickly inactivates enzymes that can degrade (break down) RNA. RNA is less stable than DNA and thus much more difficult to purify intact. **Denatured** proteins and cell debris are then removed by **centrifugation** and the cleared lysate (liquid) subjected to further purification using a **silica**-based solid support, such as a membrane or magnetic bead. Following a number of wash steps to remove impurities and contaminants, and an enzymatic step to remove the DNA, the purified RNA can be removed from the silica support using water. The RNA may also be precipitated out of the lysate using ethanol, since its like DNA, and is not soluble (dissolved) in alcohol. The removal of the contaminating genomic DNA is critical for later downstream analysis since it can greatly interfere with later manipulations of the purified RNA. Older methods of RNA purification used harsh organic chemicals, but these have been replaced with safer silica-based methods. Many biotechnology reagent companies develop and sell kits for the purification of RNA. RNA content, unlike DNA content, can vary dramatically between tissues, cell types, physiological states, and organisms. Depending on the purification method employed, either total RNA (mRNA, rRNA, and sometimes tRNA and other small RNAs), only messenger RNA alone, or both may be purified. Different applications of RNA require different input RNA molecules for optimal success. In general, most scientists are interested in mRNA, so use either purified mRNA or total RNA for their analysis.

Sample Type	Amount of Sample	Average Yield of RNA
Liver	10 milligrams	~40 micrograms
Kidney	10 milligrams	~10 micrograms
Heart	10 milligrams	~5 micrograms
Spleen	10 milligrams	~30 micrograms
Bacteria	1×10^9 cells	~80 micrograms
Tomato leaf	100 milligrams	~24 micrograms

Centrifugation

The use of centripetal or gravitational force to separate mixtures of molecules based on their relative densities; more dense components migrate farther from the axis of the centrifuge and thus move to the bottom of the tube while less dense components stay near the top in the liquid; generally measured in revolutions per minute (rpm).

Denature

To relax or unfold the structure of a biomolecule such as a protein; often causes protein to stick together in a random manner.

Detergent

A substance that can solubilize membranes or lipids.

Guanidine

A strongly alkaline crystalline compound formed by the oxidation of the nucleotide base guanine; it has strong chaotropic properties and can denature proteins.

Silica

An oxidized version of the element silicon with the chemical formula SiO_2. Silica is the most commonly found as sand or quartz in nature.

RNA Analysis

Following purification, analysis is required to determine the concentration, size, and integrity of the RNA. Methods similar to those used for DNA can also be employed for RNA, including measuring **absorbance** at 260 nanometers and agarose gel electrophoresis. By measuring the amount of light absorbed at 260nm using a **spectrophotometer**, the concentration of RNA can be calculated using the following equation: (Absorbance of the RNA value at 260nm) x (40 micrograms/milliliter) = the concentration of purified RNA in micrograms per milliliter. If a dilution was made of the purified RNA prior to the absorbance reading, then this dilution factor must be taken into consideration when calculating the final RNA concentration in the purified sample. The purity of the RNA can also be estimated by calculating the ratio of absorbance at 260nm versus the absorbance at 280nm (where proteins absorb), similar to DNA. If the ratio is >1.7 and < 2.1, the RNA is considered free of protein contamination and of high purity. **Fluorescent** methods also exist to quantify RNA, and tend to be very sensitive.

The quality and size of the purified RNA can be determined using agarose gel electrophoresis, since RNA is negatively charged like DNA (because of the sugar-phosphate backbone). Very similar techniques are used for both, except that RNA requires treatment to denature the partially folded RNA into a single-strand prior to electrophoresis. This generally involves heating the RNA in a denaturing buffer containing **formamide** before loading the RNA samples onto the gel. The denatured RNA is then separated by size, with the smaller RNA pieces migrating farther in the gel toward the cathode and the larger RNA pieces staying closer to the top of the gel. The agarose acts as a mesh to impede the migration of the RNA molecules based on their size. From a total RNA sample, the most obvious RNA molecules are the abundant 18S and 28S eukaryotic ribosomal RNA molecules from the small and large ribosomal subunits. The mRNA present in the total RNA sample will appear as a light smear throughout the entire lane of the gel, since mRNA molecules only constitute 1-3% of the total sample and they are of various sizes. Transfer RNAs and other small RNAs will migrate farthest on the gel and usually appear as a diffuse band near the bottom of the gel or closest to the cathode and below the ribosomal RNA bands. High quality RNA should demonstrate certain size characteristics and thus be useful for downstream applications.

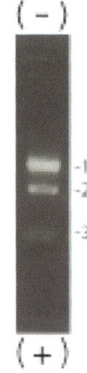

Total RNA analyzed by agarose gel electrophoresis. Note the presence of the large subunit rRNA band (1), the small subunit rRNA band (2), and transfer RNA (3).

Other RNA Analysis Methods

There are many other ways to analyze or manipulate RNA in addition to absorbance and gel electrophoresis. These include Northern Blot analysis, reverse-transcription PCR (RT-PCR), in vitro transcription, and gene expression microarrays. Northern blot analysis takes advantage of the ability to transfer separated RNA molecules from an agarose gel to a supportive membrane and then probe that membrane to determine which specific mRNAs are being made in a particular cell or tissue or physiological condition, as well as the quantity. This technique allows you to investigate the entire mRNA molecule and to determine its size and abundance.

Reverse transcription PCR (RT-PCR) also allows scientists to determine and quantify how much of a particular mRNA was present in a cell or tissue when the sample was lysed and the RNA was purified. The general process of PCR (polymerase chain reaction) basically generates millions of copies of a specific DNA or RNA sequence, which allows subsequent detection by agarose gel electrophoresis or other fluorescent methods. The ability to monitor and measure how much of a particular mRNA is made at any given time in the cell is extremely important in medical and other types of biological research. Comparing the pattern of mRNA expression between normal and diseased tissues (like cancer) may tell you something about what is causing the disease and possibly how to prevent or treat it. RT-PCR and Northern blot analysis are just two of several methods available for measuring gene (mRNA) expression.

Transcription is the process in the cell that generates RNA molecules for use in protein expression. *In vitro* transcription is the generation of RNA molecules in a test tube instead of inside a cell. This process utilizes purified RNA polymerase enzymes and a cloned piece of DNA that contains the gene of interest. The reaction also requires nucleotide building blocks (NTPs), and a buffer that maintains the proper pH and salt concentration for the RNA polymerase. The most common RNA polymerases used are from bacterial viruses (**bacteriophages**) called T7, SP6, or T3. These RNA polymerases are used because the promoter regions they recognize are short and well characterized and easy to insert before the region of DNA of interest for transcription.

Bacteriophage
A virus that infects bacteria.

Reverse Transcription
Process of making a DNA copy (complimentary DNA) from an RNA template; uses special enzymes called reverse transcriptases.

Gene expression microarrays

Gene expression microarrays are another method used to monitor and compare gene expression patterns between different RNA samples. Microarrays are specially treated glass microscope slides that have been spotted with hundreds to thousands of tiny droplets of specific short DNA sequences. These DNA sequences, which are 25-50 nucleotides in length, are permanently placed onto the slides in a very specific pattern. When the slide is incubated with an RNA sample that has been labeled with a fluorescent marker, spots that are complimentary to a specific mRNA sequence will bind to that mRNA and appear as a fluorescent spot. These spots can be detected and quantified using special machines called microarray readers. Thus, more of a particular mRNA that is present in a sample means, a higher fluorescent signal is generated. Microarrays are used to compare the gene expression patterns of two or more different samples for hundreds to thousands of different genes simultaneously. This is a very powerful technique for rapid screening of gene expression.

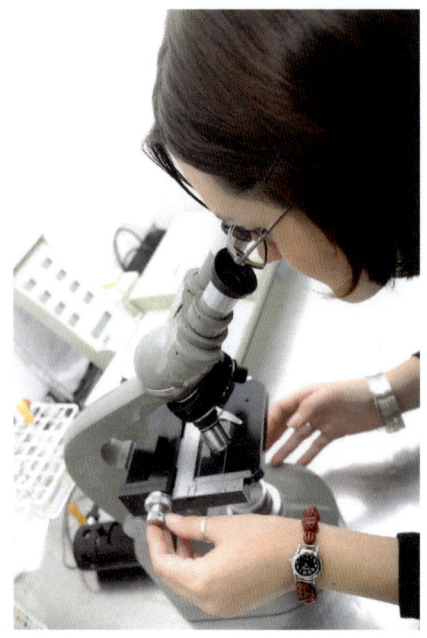

Scientist analyzing the integrity of a spotted microarray before use

RNA and Biotechnology

Although DNA is used extensively in biotechnology, RNA also plays a crucial role. Whether for monitoring gene expression patterns or generating RNA using in vitro transcription for in vitro protein expression, many questions in biotechnology can be answered only by analyzing RNA. Many new techniques being developed in biotechnology deal with the analysis of very small amounts of RNA from very precious and rare samples. These techniques enable the investigation of which genes (and, hopefully, which proteins) are involved in various disease states, such as cancer, diabetes, autoimmune diseases, infectious diseases, and many others, supporting additional drug development in these areas. Moreover, new mechanisms for the regulation of gene expression have been identified recently that may revolutionize how some diseases are treated. A class of small RNAs (called small interfering RNAs or siRNAs) appear to regulate the expression levels and patterns of many genes in development, normal growth and homeostasis, and in diseases. This discovery of RNA interference has opened up a whole new field of biology, also highlighting the importance of RNA within the cell.

Summary

RNA purification is a classic method used in biotechnology. Many different methods are available for purifying RNA from the other components of cells and tissues, but most often takes advantage of the negative charge on the RNA to bind to silica solid supports. The purified RNA can be analyzed using different methods to determine its quality and quantity, but also the roles of RNA in disease and treatments..

Concept Reinforcement

1. Describe the RNA purification process using a silica-based membrane method.

2. Would the presence of DNA in an RNA sample interfere with absorbance and quantitation analysis?

3. When analyzing total RNA by agarose gel electrophoresis, what do you expect to see?

Section 2.12 – Chemistry of Proteins

Section Objectives

- Describe the basic chemical structure of a protein

- Describe the process of protein synthesis called translation

Chemistry Structure of Proteins

Proteins are large organic compounds composed of strings of amino acids that are covalently linked through peptide bonds between adjacent **amino acids**. The carboxyl group from one amino acid is permanently joined to the amino group of the next amino acid. The sequence of amino acids is determined by the sequence of nucleotides in the DNA gene and resulting messenger RNA. There are 20 standard amino acids used in protein synthesis, although other nonstandard amino acids are used by some organisms or are generated following translation. Each amino acid has a unique side chain that has different chemical properties and produces different structures when combined in unique ways. Proteins play extremely important roles in cells and organisms. They provide the physical structure for cells and also allow them to perform metabolism, replication, and communication within the cell or between neighboring or distant cells. The unique and essential properties of proteins make them indispensable for life!

> ### Amino Acid
> The basic building blocks of proteins; 20 standard amino acids are used in protein synthesis.

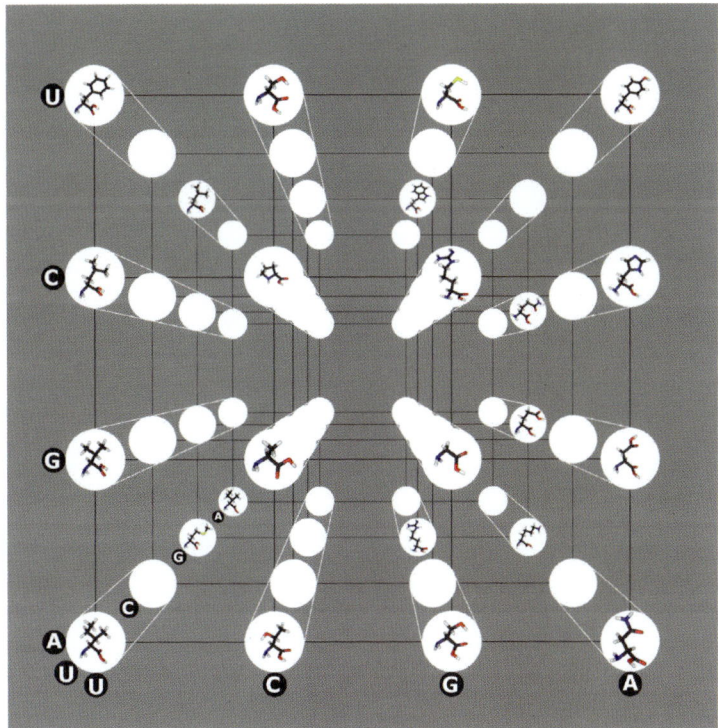

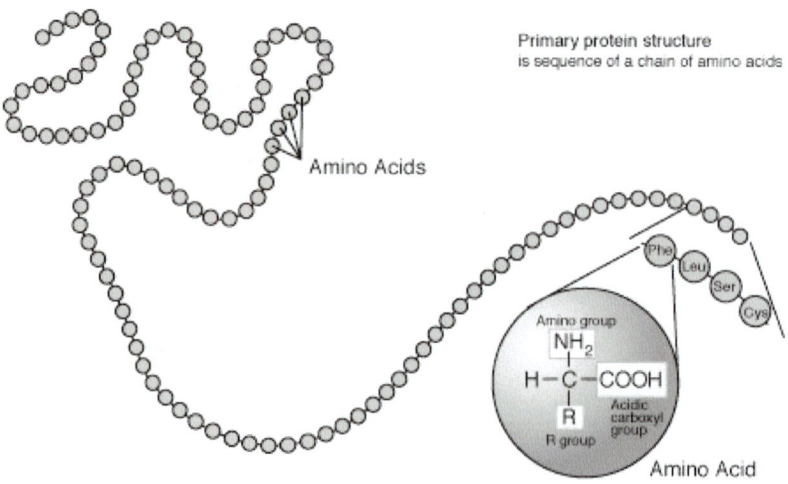

Primary protein structure
is sequence of a chain of amino acids

Amino Acids

Amino group
NH_2

$H-C-COOH$

R

R group

Acidic
carboxyl
group

Amino Acid

Polypeptide (protein) composed of a sequence of amino acids
linked together by peptide bonds

Translation (Protein Synthesis)

As referred to earlier, proteins are synthesized in the cell through a process called translation that occurs in the cytoplasm of prokaryotes, eukaryotes, and archaea. All three main subdivisions of life on earth undergo translation using similar mechanisms, though there are distinct differences between them. This section will focus on prokaryotic translation, also highlighting differences regarding translation in eukaryotic cells. Both processes require numerous protein and ribosomal RNA complexes for successful protein production. Many of the proteins involved are **GTP**-binding proteins that bind to the nucleotide GTP and use energy from GTP hydrolysis (break down) to carry out a particular step in translation.

> **GTP: Guanosine-triphosphate**
>
> GTP is used as an energy source in translation and as a building block of RNA.

Translation can be divided into four steps:

1. activation

2. initiation

3. elongation

4. termination

Technically, activation is not a step in the translation process, but is required to prepare the charged, amino acid carrying, transfer RNAs necessary for elongation.

Initiation

In the first step of translation, the 5' or beginning end of the messenger RNA (mRNA) is recognized by the small ribosomal subunit with the aid of three different initiation factors. Prokaryotic and eukaryotic initiation factors recognize different features of the

154

mRNA to start translation. In prokaryotic cells, a specific sequence in the mRNA called the Shine-Delgarno sequence is recognized and binds to the complimentary sequence in the RNA component of the small ribosomal subunit. In eukaryotic cells, the 5'-cap and 3'-polyA tail are recognized by the initiation factors to begin assembly of the translation complex and more factors are involved, making eukaryotic translation initiation much more complex. The first amino acid (**methionine**) is then brought to the complex and positioned into the "P" or peptide site. The large subunit of the **ribosome** then binds and causes release of the initiation factors.

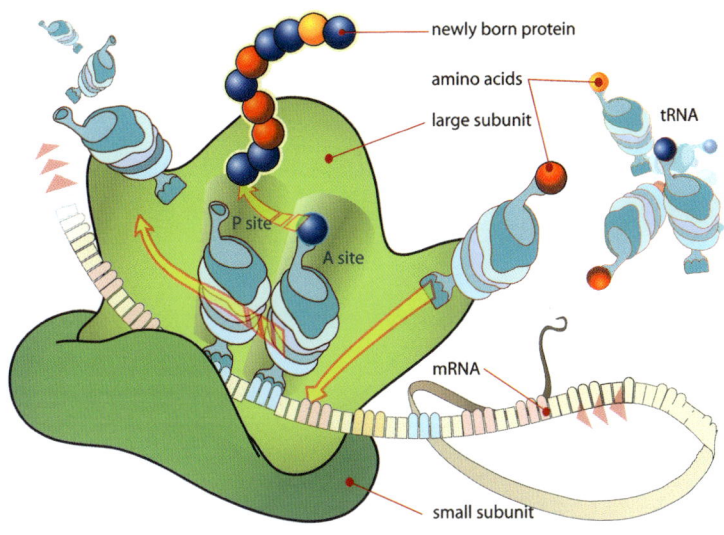

Diagram illustrating the process of translation

Elongation

During elongation, amino acids are positioned at the "A", or amino acid site, in the ribosome by charged tRNAs and a peptide bond is formed between the amino acids in the "A" and "P" sites (**transpeptidation**). Then the growing peptide chain is moved to the "P" site, freeing up the "A" site for the next amino acid in the mRNA sequence (**translocation**). This process involves elongation factors and the hydrolysis of GTP to provide the energy needed. In addition, there is a proof-reading function that makes sure the correct tRNAs are matching with the mRNA sequence, to ensure the correct amino acids are incorporated into the growing protein chain.

Termination

The final step in translation is the termination step. This occurs when the ribosome recognizes a stop codon in the "A" site in the mRNA template. This codon does not code for an amino acid, but instead signals various termination factors to bind to the complex and to release the completed protein. In addition, the termination factors signal the translation complex to fall apart. Eukaryotic cells have only one termination factor while prokaryotic cells have three (one for each of the three different stop codons). Following the release of the protein from the translational machinery, the protein can complete the folding to its final shape, though often this occurs during translation.

Methionine

The first amino acid incorporated into a protein.

Ribosome

Large protein: RNA complexes that are responsible for protein synthesis.

Elongation

The phase in translation when new amino acids are being added to the growing peptide chain.

Translocation

The movement of the growing peptide chain from the "A" site to the "P" site during translation.

Transpeptidation

The peptide bond formation between two amino acids in the "A" and "P" sites during translation.

Termination

The final step in translation when completed protein is released and the translation machinery fall apart.

Structure of Proteins

The size of a protein can be measured by the number of amino acids it contains and by its total molecular weight, which is normally referred to in units of **Daltons** (Da; atomic mass units) or **kilodaltons** (kDa). Proteins can range in size anywhere from only a few amino acids (peptides, several Da), to thousands of amino acids (hundreds to thousands of kDa). Each protein folds into a unique three-dimensional shape, often called its native shape. There are proteins in the cell which can help other proteins fold properly or find the right binding partners, and these proteins are called **chaperones**. The structures of proteins are divided into primary, secondary, tertiary, and sometimes quaternary:

Primary structure: the amino acid sequence.

Secondary structure: regularly repeating structures stabilized by local hydrogen bonds; examples include alpha helices and beta pleated sheets.

Tertiary structure: the overall three-dimensional shape of the protein stabilized by nonlocal interactions through hydrogen bonds, salt bridges, disulfide bonds, and hydrophobic interactions.

Quaternary structure: the shape or structure of the protein when interacting with other proteins or biomolecules in a complex.

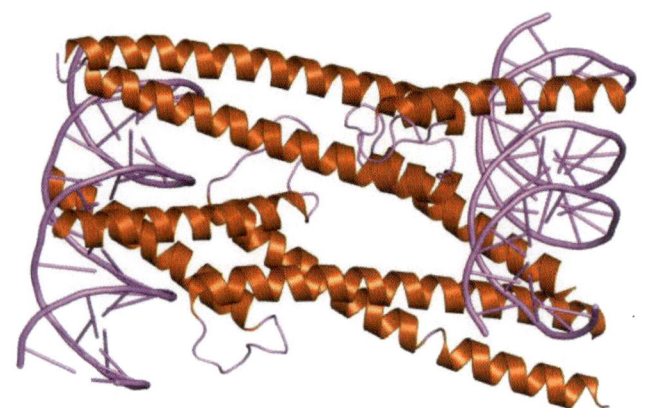

Example of alpha helices in a protein

Proteins are not rigid molecules, but can shift between related structures. This influences their biological functions. They are usually divided into three main tertiary shapes: 1) globular, 2) fibrous, or 3) membrane. Globular proteins are generally soluble in the cytoplasm, while fibrous proteins often provide structural support to the cell or organism. Membrane proteins are found embedded in or associated with the cell membrane and often serve as receptors on the surface of the cell to receive signals from other cells or other parts of the organism. Some may act as channels to allow certain molecules to pass into or out of the cell.

Following translation, proteins may be further altered by various post-translational modifications. These generally involve the addition of other types of molecules to amino acids

---

Daltons (Da)

The unit of measure for protein mass.

Kilodaltons (kDa)

One thousand Daltons.

Chaperones

A protein or group of proteins that can assist newly-made proteins to fold correctly.

156

in the target protein, such as carbohydrates (sugars), lipids, or phosphate groups. These modifications usually affect the activity or function of the protein dramatically. Often, they are involved in making sure the protein gets to the right place in the cell.

Summary

Proteins serve many different functions in cells and tissues and are composed of strings of amino acids joined together during a process called translation. The mRNA template generated during transcription is then used as the template during translation where ribosomes and other factors assemble on the RNA to direct the correct linkage of amino acids to form mature proteins. Proteins play an incredibly important role in biotechnology.

Concept Reinforcement

1. How many standard amino acids are there and how are they different?

2. How can the different side chains on the amino acids affect the shape and function of a protein?

3. What are the three main steps in translation? What is occurring at each step and what factors are involved?

4. Name two differences between eukaryotic cells and prokaryotic cells concerning translation.

Section 2.13 – Protein Functions

Section Objective

- Explain the differences between types of proteins and their functions within a cell

Role of Proteins within Cells

Proteins play key roles in all cellular processes and all events that occur within an entire organism. Some proteins are **enzymes** that catalyze biochemical reactions in the cell to allow for energy production and metabolism. There are at least 20 different enzymes involved in **glycolysis** and the **Kreb's cycle**, both key pathways in energy production from carbohydrate sources such as glucose or glycogen or from fats. All of the important biomolecules in the cell, like DNA, RNA, carbohydrates, lipids, and other proteins could not be produced without many different enzymes and cofactors. Degradative (destructive) enzymes help break down biomolecules in the cell that it no longer needs and can recycle the building blocks for new biomolecules.

> ### Enzyme
> A particular class of proteins; an enzyme catalyzes a specific biochemical reaction.
>
> ### Glycolysis
> The enzymatic pathway that generates energy from glucose.
>
> ### Kreb's cycle
> The enzymatic pathway that produces energy, starting with the products of glycolysis and leading to the production of intermediates that feed into the electron transport system.

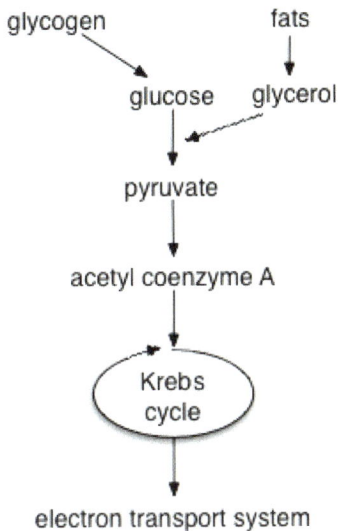

Simple schematic of the energy production cycles involved in glycogen, glucose, and fat metabolism in the cell. All of these steps involve specific protein enzymes.

Other proteins act as structural components of the cell to provide strength and may even allow for movement. For example, actin and myosin form part of the **cytoskeleton** in muscle cells. Many proteins in the cell are involved in sending signals within the cell or between neighboring or distant cells, or tissues, that tell the cell to grow, release hormones, migrate, or even to die. These signals are often linked in complex patterns called signal transduction pathways.

Cytoskeleton

The supporting framework or scaffold within the cell that is composed of various sized filaments. The cytoskeleton facilitates many functions and activities in the cell, including cellular motion, intracellular transport, and cell division.

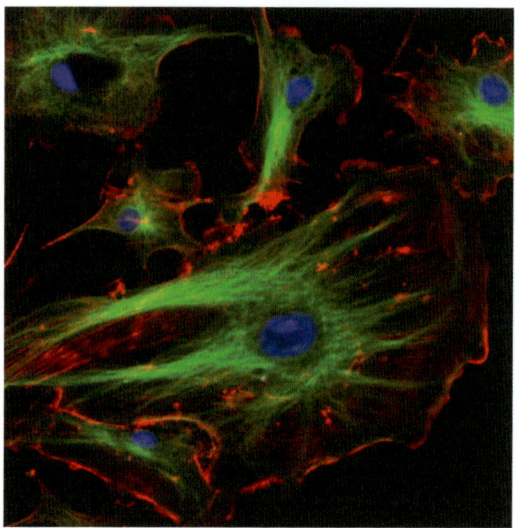

Fluorescent labeling of various components of the cytoskeleton. Actin is pictured in green, while the nucleus appears blue and the cell membrane appears red.

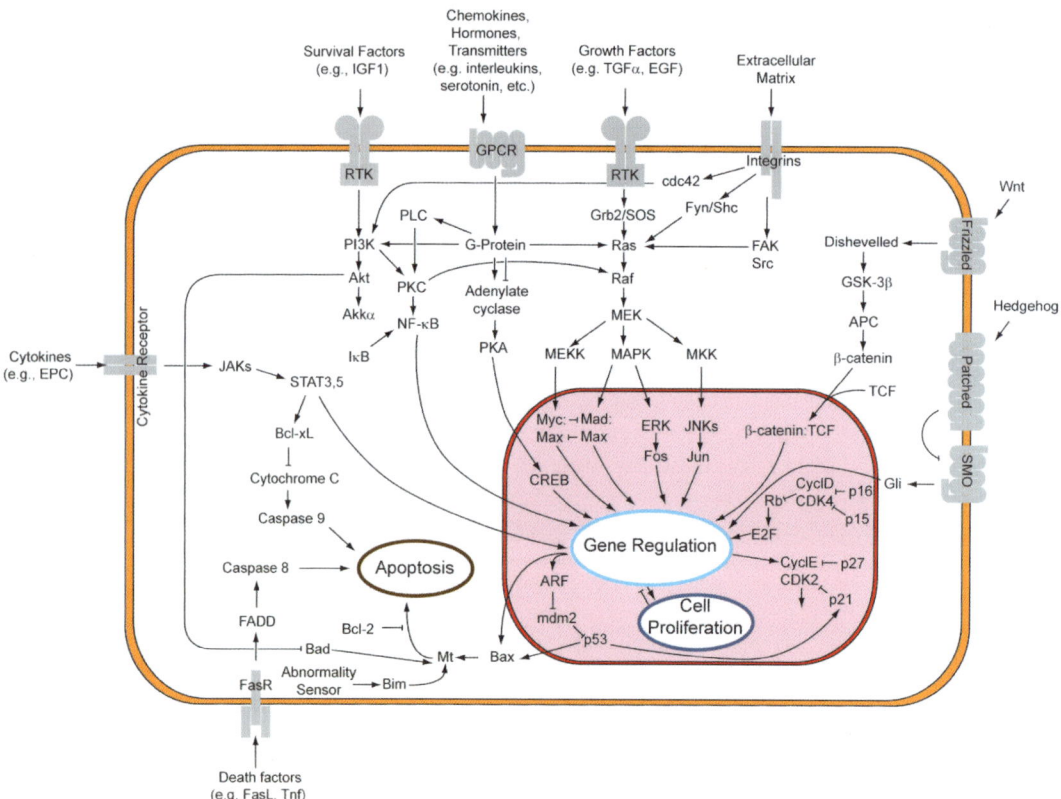

Typical signal transduction pathways in a eukaryotic cell. Each point indicates a different protein component that leads from signals outside the cell to points in the nucleus that tell the cell to proliferate (grow).

Many systems in our bodies could not operate properly without proteins. An example would be the immune system, which relies on numerous protein receptors on the surface of cells that recognize different protein hormones and growth factors to mount an appropriate immune response to bacteria or viral infections. Part of that immune response might be the production of protein **antibodies** to help fight the infection.

160

Structure Versus Function

The amino acid sequence of a protein determines what the final structure will be, through the complex interaction of amino acids with each other in space. Amino acids that are distant from each other in the primary amino acid sequence may be very close together in the final folded protein. Although each protein generally has a unique function, many proteins have similar functions that can be grouped together into families. These may include proteins that provide structural components to the cell as they tend to be fibrous in shape with common secondary structures. They might be proteins with similar enzymatic activities. One such example would be a family of proteins called kinases. A **kinase** is an enzyme that transfers a phosphate group from **ATP** to another protein or lipid molecule, but usually the target is a protein. Phosphorylation can alter the activity, location, or stability of a protein and thus can dramatically affect the physiology of the cell. Many kinases are involved in signaling pathways in the cells that control cell growth, cell death, and other key cellular events. It is estimated that there are at least 518 different kinase enzymes in a human cell. Although each kinase protein recognizes a distinct target or a small number of targets, they almost all have a series of conserved amino acids that come together in the final folded protein to form an ATP binding domain or pocket. All kinase enzymes require ATP as the source of phosphate to transfer to the target, so all must be able to bind to ATP. They accomplish this by all containing a similar structural motif or feature in them. These types of motifs that are responsible for specific functions are common in the protein world and are often called protein **domains**.

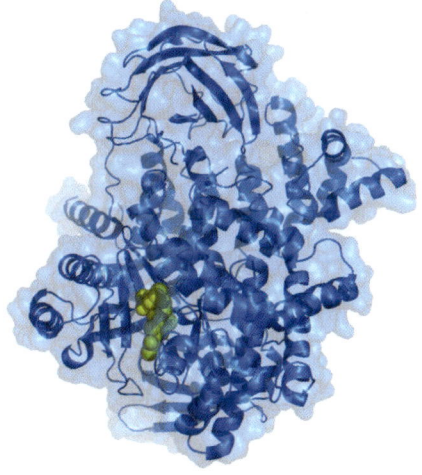

Kinase enzyme showing the ATP bound in yellow

Summary

Proteins are indispensable to every cell and organism. They perform such vital functions as providing the structure of the cell, the ability to generate energy and perform various chemical reactions, the ability to copy the DNA for cell division and growth, and the ability of the cell to send signals to itself, neighbor cells, and even distant cells within the organism. Thus, proteins have unique structures which allow them to accomplish so many different functions within a cell or organism.

Antibody

A protein produced by the immune system to recognize foreign antigens and bind to them to fight viral and bacterial infections.

ATP Adenosine triphosphate

A source of phosphate for kinase reactions. ATP is also used as a nucleotide building block for RNA synthesis and source of energy in biochemical reactions.

Domains

A specific region of a protein that has a specific function; often found in proteins with very similar functions.

Kinase

An enzyme that transfers a phosphate group form ATP to a target molecules, usually another protein.

Concept Reinforcement

1. Name a few different functions that proteins serve in a cell and why they are important to the cell.

2. What function does the cytoskeleton serve in a cell?

3. What is meant by a protein domain or motif?

Section 2.14 – Protein Expression and Purification

Section Objectives

- Discuss methods that can be used to express proteins in cells (*in vivo*) or in a test tube (*in vitro*)

- Describe methods for protein purification

Protein Expression Methods

In Vivo

Proteins can be expressed and purified from cells, tissues, or organisms where they are normally produced. This is how classical biochemistry was done and, although successful, it usually resulted in very low yields of the purified target protein making them very difficult to study. In fact, each cell makes thousands of different proteins at any one time and no single protein is in a large amount, unless it is a very common cytoskeletal protein or ribosomal protein. Most of the proteins in a cell that have unique and critical signaling functions are present in very low amounts.

With the advent of molecular biology and the ability to obtain the gene for any protein and clone (transfer) that gene into another organism, protein expression took on a new proportion. Now bacteria, yeast, plants, or other organisms could be engineered to express the target protein, often at high levels, allowing for efficient expression and subsequent purification and study. Bacteria are the most common source for target protein expression because they are easy to manipulate and inexpensive to grow in large quantities. Yeast, insect, and mammalian cells are also used for this purpose, but are more complex and expensive to work with. Depending on the characteristics of the target protein, one of these expression systems might be selected. In general, human proteins are best expressed in mammalian cells since they are the most similar in their translational machinery and protein processing. Once the target protein is expressed, it can then be purified using several methods. These *in vivo* expression and purification systems can often produce **milligrams**, **grams**, and even **kilograms** of the target protein, depending on the number of cells grown. This scale of protein production may be enough for large experiments or even enough of a biotherapeutic protein for treating patients.

In Vitro

The proteins expressed *in vitro* are usually done using one of three commercially available expression systems and are often called "cell-free" expression systems. A number of companies have *in vitro* translation systems available and include Promega Corporation, Roche Biochemicals, Ambion, and Novagen (EMD Biosciences). These are biotechnology companies that develop, manufacture, and sell products used in basic research, both medical and agricultural, in drug discovery, in the development of diagnostic tests, and in the purification and analysis of biomolecules such as DNA, RNA, and proteins.

In vivo

Inside an intact cell or organism.

Grams

The base unit of measure for mass.

Kilograms

One thousand grams.

Milligrams

One thousandth of a gram (0.001 gram).

In vitro

"In glass." Occuring outside of a whole cell. Carried out in a test tube.

Endogenous

Normally present within a cell.

Exogenous

Introduced from outside of the cell. Not normally found in that cell type.

All of the *in vitro* expression systems are complex lysates made from either mammalian cells, bacterial cells, or plant extracts in which the cells have been lysed (broken open) and the contents released. These lysates contain all of the necessary components for translation to occur that are either present naturally in the lysate or are supplemented into the lysate preparation. This includes ribosomes, translation initiation, elongation, and termination factors, GTP and ATP for energy, charged tRNAs, amino acids, the correct salts, and a buffer to maintain the proper pH. In general, a scientist only needs to add mRNA to translate a target protein. However, for this to occur the **endogenous** mRNA in the lysate had to first be degraded so that only **exogenous** mRNA added is translated.

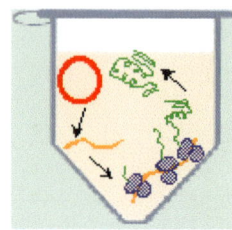

Diagram illustrating *in vitro* protein expression from a plasmid template that contains the gene for the target protein, showing transcription and translation.

The mRNA added to cell-free expression systems that will act as the guide for translation is generated using a process called *in vitro* **transcription**. This allows specific mRNA molecules to be made using a purified RNA polymerase enzyme and RNA building blocks (ribonucleotides), with a DNA molecule as the template. The DNA template is usually a **plasmid** in which the gene of interest (that codes for the target protein) has been inserted using recombinant DNA techniques, such as PCR or restriction enzymes.

Mammalian Cell Expression System

One of the most common cell-free expression systems is based on a lysate made from rabbit reticulocytes (red blood cells). These cells contain all of the necessary machinery for protein synthesis, but lack nuclei, and only contain mRNA for globin, the protein necessary for oxygen transport in the blood. The globin mRNA is removed using an enzyme that degrades RNA. This enzyme is then inactivated to allow for selective protein expression of the exogenous RNA added to the lysate.

The rabbit reticulocyte lysate system with intact globin mRNA is often used to study the process of translation itself or factors that might affect translation, such as potential **antibiotics**, since they often preferentially inhibit bacterial translation and not mammalian translation. Proteins produced from added mRNA are synthesized almost as efficiently as the normal globin mRNA, and makes this a very robust system for cell-free protein expression. The rabbit reticulocyte lysate expression system is particularly suited for mammalian proteins.

Plant Cell Expression System

A second cell-free expression system is based on an extract made from wheat germ. The germ is the heart, or embryo, of the seed kernel and is very active metabolically. Thus, it contains the active translation machinery necessary for making proteins. Wheat germ extract tends to have lower levels of background translation due to the low levels of endogenous RNA in the extract, providing an advantage over rabbit reticulocyte lysate. Proteins from many different organisms, including human, mouse, rat, plant, yeast, and bacteria, are effectively synthesized using wheat germ extract.

Bacterial Cell Expression System

Another cell-free protein expression system is based on a crude bacterial cell lysate called S30. Different strains of *E. coli* bacteria are used to prepare the lysate. Because the extract contains high amounts of endogenous mRNA, it is incubated during the manufacturing phase to allow all of the host mRNA to be translated. Then the mRNA is rapidly degraded by the lysate itself. While the rabbit reticulocytelysate system is fairly complex, the bacterial S30 system is less complicated and thus very efficient at translation. Due to the ability of bacterial extract to degrade RNA, the template for translation must be DNA. This requires the gene for the target protein be in a plasmid or PCR product that also contains the appropriate bacterial promoter region for transcription to occur.

Fluorescent

A chemical or compound that absorbs light at a particular wavelength and then emits light at a lower wavelength.

All of the three mentioned cell-free protein synthesis systems have the advantage of being able to add modified or unnatural amino acids for incorporation into the target protein. This can significantly aid in analysis of the protein, by making it easier to detect or purify. One example of a modified amino acid that can be incorporated into a protein is a fluorescently labeled amino acid that contains a **fluorescent** molecule attached to a lysine amino acid. Fluorescent detection is very sensitive and allows for easy analysis of the expressed (and fluorescently-labeled) protein.

Protein Purification Methods

The study of proteins usually requires that they first be purified in some manner to allow detailed study of their structures and functions. Protein purification takes advantage of unique features of the protein of interest versus all of the other proteins and biomolecules in a cell. Unlike DNA or RNA (highly negatively charged, very similar to each other, and thus relatively easy to purify), each protein within a cell is unique and may require a unique purification scheme to be successful. The unique feature may be the protein size, charge, shape, enzymatic activity or function, or location within the cells. Most likely, a combination of these features is utilized to purify a protein. Protein purification is often accomplished using a technique called **chromatography**. Proteins may be purified inside the cell, from the cell membrane, from the surrounding fluid if the protein is secreted from the cell, or from a cell-free expression system. The enzymes or therapeutic proteins purified from bacteria must be highly pure, especially in the case of proteins that will ultimately be injected into humans or animals.

Chromatography

The physical separation of a mixture of compounds by using a mobile and a stationary phase; often used with proteins and other biomolecules for detection and purification.

The first step in protein purification is to gain access to the protein of interest by lysing the cell that it is expressed in, unless the protein is secreted from the cell. The protein is then bound to some type of solid support that will allow many other proteins and contaminants to be washed away. The target protein is then removed from the solid support, usually by some type of buffer change, for further purification with a different solid support. The solid support may be a membrane or small spherical beads or particles that allow a specific feature of the protein to bind to a surface, while excluding most other proteins and cellular molecules (i.e. DNA, RNA, sugars, and lipids). To aid in purification, some proteins can be altered at the genetic level to add a "tag" to them that makes purification easier. These tags are usually relatively short amino acid sequences that can bind specifically to a chromatography matrix, while all other proteins and contaminants can be removed.

As noted earlier, proteins are the heart and soul of biotechnology. They are enzymes required for numerous industrial processes, basic research, as well as biotherapeutic drugs utilized to treat human and animal diseases.

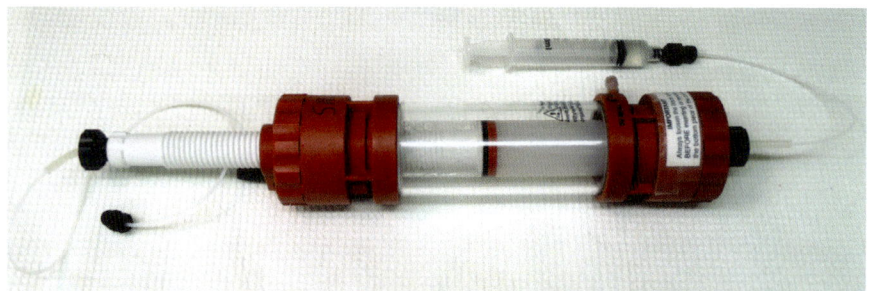

Typical chromatography column for protein purification

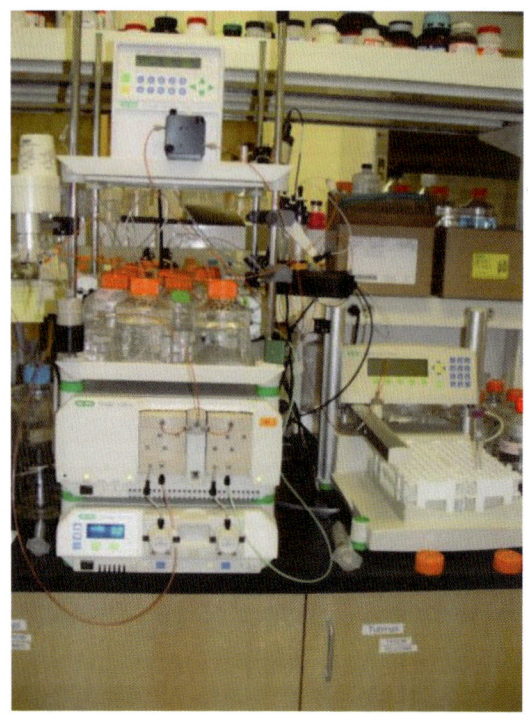

Chromatography system for protein purification, including pumps, chromatography column, and fraction collector

166

Summary

Proteins are expressed in every living cell and organism, but can also be expressed in a test tube using a process called *in vitro* translation. This system uses complex biological mixtures from lysed animal, plant, or bacterial cells that contain all of the necessary components for protein synthesis. Organisms may also be engineered to express proteins. This is very commonly done with bacteria because they make good protein expressing factories. Once a protein is expressed, it is purified away from all of the other cellular constituents using chromatography, or serial removal, of unwanted biomolecules, leaving only the protein of interest. Because each protein is unqiue, it may involve a unique and individualized protein purification scheme. The expression and purification of specific proteins is key to many different uses of biotechnology in our everyday lives.

Concept Reinforcement

1. What is the biggest problem with expressing and purifying proteins from their native source?

2. When might you use an *in vivo* expression system for a protein and when might you choose an *in vitro* system?

3. What does chromatography mean and how does it apply to protein purification?

Section 2.15 – Protein Analysis

Section Objective

- Explain various techniques used to analyze proteins

Analysis of Protein

Once a protein of interest has been expressed, either *in vivo* or *in vitro*, protein purification is most often the next step in the process of studying the protein thoroughly. Following purification, the next step is analysis of the purified protein, for which there are numerous methods available. These include methods to determine the structure, size, shape, identity, or activity/function of the protein.

Gel Electrophoresis

Size determination is usually accomplished using gel electrophoresis. The protein is separated through a chemical matrix or gel using electricity. Smaller proteins move more quickly through the matrix than larger ones. It is simply harder for the latter to move through the meshwork matrix of the gel. The technique of polyacrylamide gel electrophoresis (**PAGE**) is commonly used in basic research and biotechnology and is very similar to agarose gel electrophoresis used for DNA and RNA analysis. Instead of agarose, it uses a different matrix called polyacrylamide, which allows for better separation of molecules that are much smaller than RNA or DNA, such as proteins. Once the proteins are separated in the gel matrix they are usually visualized using some type of stain, such as Coomassie Blue.

> **PAGE**
> Polyacrylamide gel electrophoresis. A process used to separate proteins by size.

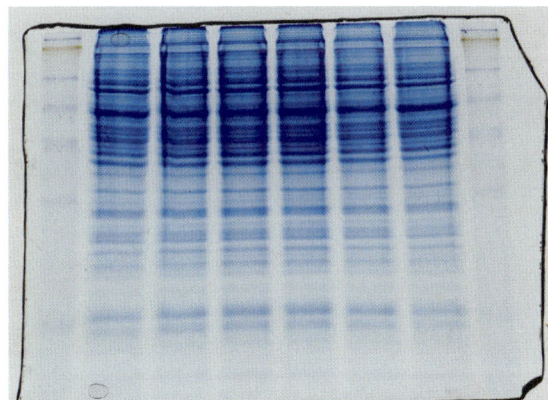

PAGE gel showing different proteins separated by size and then visualized by staining with Coomassie Blue. Lanes 1 and 8 contain a protein molecular weight marker of known sized proteins.

Mass Spectrometry

A method to determine the mass of a biological molecule, such as a peptide or protein, by inducing a charge onto the molecule and then sending it down a path using magnetic or electrical direction to a detector that can determine mass.

Mass Spectrometry

Size may also be determined using **mass spectrometry** (mass spec or MS). This technique is better suited for small proteins and peptides versus large intact proteins. In mass spectrometry, the mass-to-charge ratio of a molecule is determined by first ionizing the molecule (i.e. giving it a net charge), then sending the charged molecule through a vacuum tube in the presence of a magnetic and/or electrical field in which the various sized molecules will migrate at different degrees and thus be separated based on charge. The molecules are then captured by a detector that can measure the charged molecules and determine their exact mass. There are many different modifications of basic mass spectrometry that allow for the mass determination of mixtures of peptides and proteins, as well as other types of biomolecules.

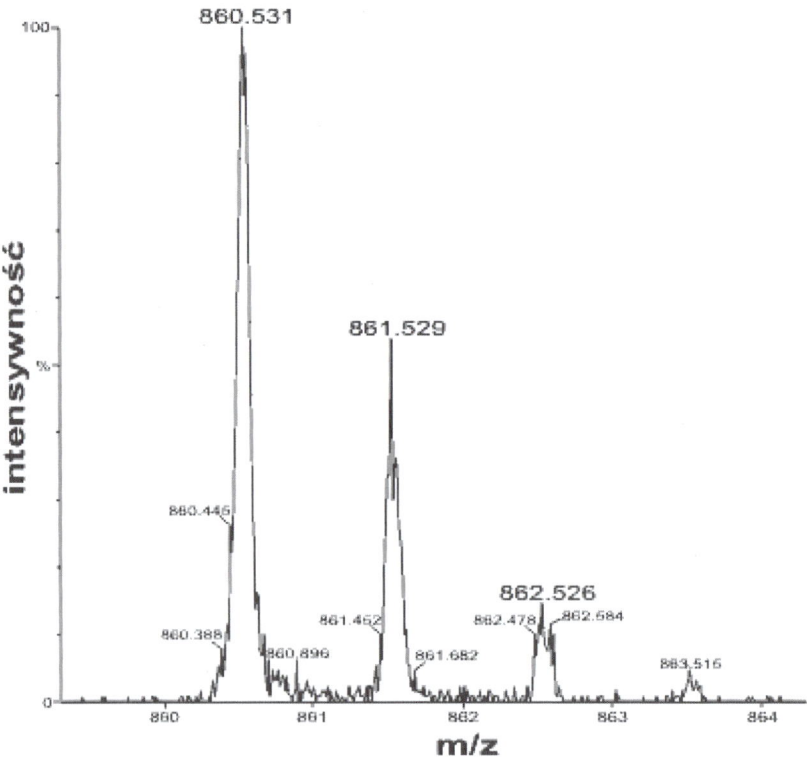

Mass spectrum of a peptide showing the isotopic distribution

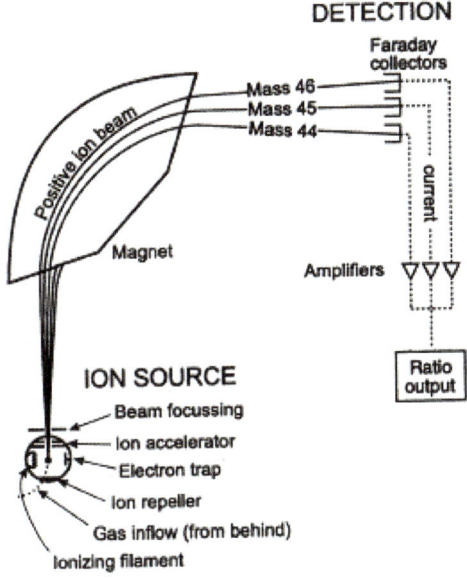

Mass Spectrometer Schematics

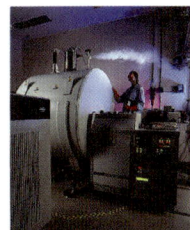

Mass Spectrometer

X-ray Crystallography

The structure of a protein is best determined using either **X-ray crystallography** or NMR (nuclear magnetic resonance) analysis, although mass spectrometry analysis may also be used. X-ray crystallography was first developed in the early 1920s. It is a method for determining the arrangement of atoms in a molecule by bombarding a crystal of the substance with X-rays and seeing the diffraction pattern that is produced. An electron density map is produced that allows the three-dimensional shape of the molecule to be determined. X-ray crystallography was first applied to proteins in the 1950s when Max Perutz and Sir John Cowdery Kendrew determined the structure of myoglobin, a muscle protein involved in oxygen transport.

The structure of over 39,000 biological molecules has been determined using X-ray crystallography to date. This will continue to be a very powerful tool in many fields of science, including biotechnology, since knowing the structure of various protein drug targets is enhancing the development of new treatments.

X-ray crystallography

A method to determine the structure of a biological molecule, such as a protein, that is based on the diffraction pattern of electrons produced after bombarding a crystal of the molecule with X-rays.

Crystal of the protein DNase, suitable for X-ray crystallography analysis

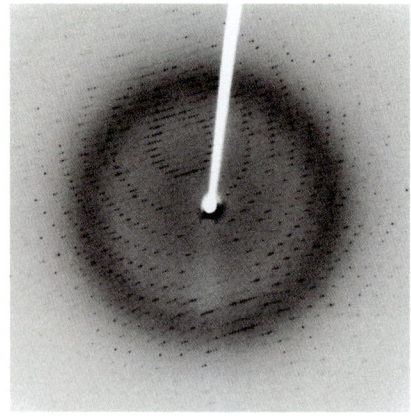

X-ray crystallography diffraction pattern of a crystallized protein

Nuclear Magnetic Resonance (NMR)

Nuclear Magnetic Resonance is another method for determining the structure or shape of a molecule. This method is based on the magnetic properties of the nuclei of the atoms in a molecule. All nuclei that contain odd numbers of protons or neutrons demonstrate quantum mechanical properties (i.e. spin) when subjected to a strong magnetic field. These properties can be measured by aligning the nuclei with a constant magnetic field, and then perturbing this alignment with an alternating magnetic field. The most commonly measured elements within a molecule are hydrogen and carbon-13, but others such as nitrogen, fluorine, phosphorus, oxygen, sodium, chlorine, and platinum can also be observed. NMR spectroscopy is generally limited to smaller molecules and proteins and has been used to determine the structure of some 6000 biological molecules. As with mass spectrometry, there are many different variations of NMR that allow for the structural determination of different types of biomolecules.

NMR spectrometer

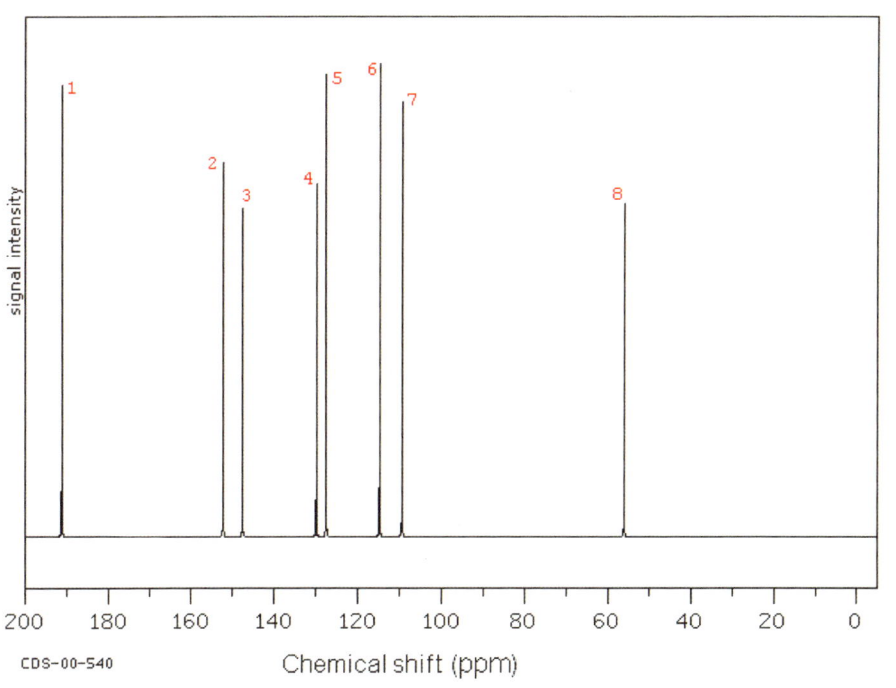

ppm	Int.	Location
191.21	955	1
152.18	791	2
147.50	692	3
129.77	746	4
127.49	975	5
114.75	1000	6 *
109.34	920	7 *
56.10	706	8

NMR spectrum (13-C) for the chemical molecule vanillin

NMR (Nuclear Magnetic Resonance)

A technique based on the changes in spin of neutrons and protons present in certain elements in the presence of a magnetic field.

In addition to utilizing NMR to analyze the structure and function of some proteins, these features can also be predicted by comparing their amino acid sequence to proteins of known function and structure. This is accomplished using **bioinformatics** and special computer software. Many types of enzymes have unique features in their amino acid sequences that account for a particular function, such as binding to a particular type of molecule, like ATP and **kinase** enzymes, or signaling proteins that have **motifs** that interact with each other to transfer signals from outside the cell to inside the cell.

The identity of a protein may be deduced from its size, but more specifically can be determined using an **antibody** that specifically reacts and binds to a particular protein or protein family. Antibodies are important proteins produced by the immune system in response to exposure to foreign molecules (**antigens**), such as in a bacterial or viral infection. They can very specifically recognize a specific part of a molecule such as a protein. Antibodies can be used in several techniques to detect and analyze proteins, the most common of which are **Western Blotting** analysis and **ELISA** assays (Enzyme Linked Immunosorbent Assay).

Term	Definition
Kinase	An enzyme family that bind to ATP and transfer a phosphate group to a different molecule, usually a protein.
Motifs	A short sequence of amino acids in a protein that is responsible for a particular function.
Antibody	The proteins expressed by specialized cells in the immune system in response to a foreign antigen; each antibody is specific to a particular antigen.
Antigen	A molecule not normally present in an organism that can stimulate an immune response.
Western Blot	A method to detect and potentially quantitate proteins that utilize antibodies specific to the target protein.
ELISA	Enzyme-linked immunosorbent assay. A method for the detection and quantitation of proteins based on the use of antibodies specific for the target protein.

Western Blot Analysis

In Western Blot analysis, the target proteins are separated by size using polyacrylamide gel electrophoresis, and then the separated proteins are transferred to a solid support membrane. The membrane is incubated with an antibody specific for the target protein and a second antibody that can bind to the first antibody. This secondary antibody is labeled with some type of enzyme or label that allows for detection by color, radioactivity, light, or fluorescence. This method allows for the detection of specific proteins, whose levels might change in a normal versus tumor cell or tissue and this information could be useful in cancer research.

Bioinformatics

A combination of mathematics, computer science algorithms, and often statistics, to answer complex biological problems, usually through data analysis.

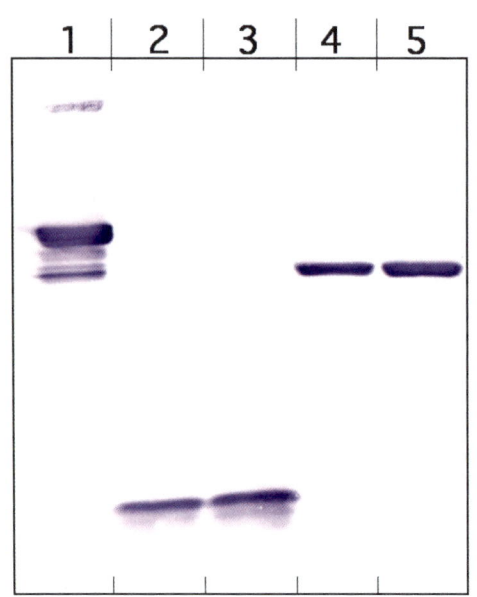

Western blot showing target proteins shown in blue are bound by a specific primary antibody and labeled secondary antibody, which binds to the first antibody.

ELISA

An ELISA (Enzyme Linked ImmunoSorbent Assay) also takes advantage of the specific nature of antibodies and their ability to recognize and bind to specific proteins. Instead of analyzing proteins separated in a gel and then transferred to a membrane first, an ELISA allows for the detection of proteins in solution in a tube or well of a multi-well plate. The proteins are then recognized in their more native (natural) state, which might be very important for the research question being asked. For example, a biotechnology company might want to know how much of a particular functional protein, such as a hormone like insulin, is present in a purification process, and could use an ELISA to measure the amount of properly-folded insulin present.

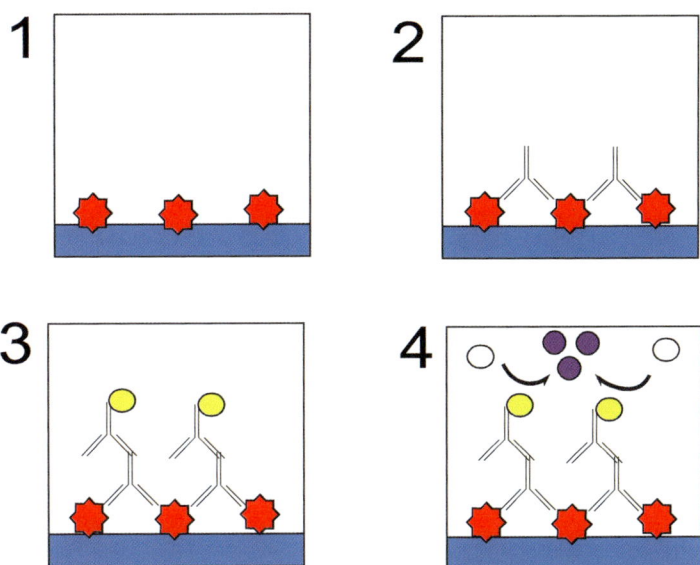

Schematic illustrating the steps in an ELISA. The target protein binds to the well of the assay plate (red star; 1). The primary antibody that recognizes the target protein then binds (y-shaped molecule; 2). The secondary antibody then binds to the first antibody (y-shaped molecule with yellow circle attached; 3). The secondary antibody is labeled with an enzyme that can convert a chemical to a specific color if present (white to purple circles; 4).

Protein Function or Activity

The function or activity of a particular protein may be difficult to determine or detect. Since proteins are unique and may have very different specialized functions, scientists may have to develop an assay to detect the activity for each individual protein. This can be quite difficult in some cases, though some proteins tend to fall in families with similar, though generally not identical, functions. Most activity or functional assays incorporate a substrate for the activity and then a way to measure the outcome through the production of a signal that can be measured using specialized equipment. These signals may be a color change, production of light, **fluorescence**, or radioactivity. Ultimately, the activity of many industrial and therapeutic proteins will need to be verified since they are part of an industrial process or drug treatment. This highlights the importance of being able to determine and measure the activity of a particular protein.

Fluorescence

The excitation of atoms in a molecule by a specific wavelength of light that results in the release of energy from that molecule at a lower wavelength of light. The light released can be detected using special cameras, microscopes, or readers.

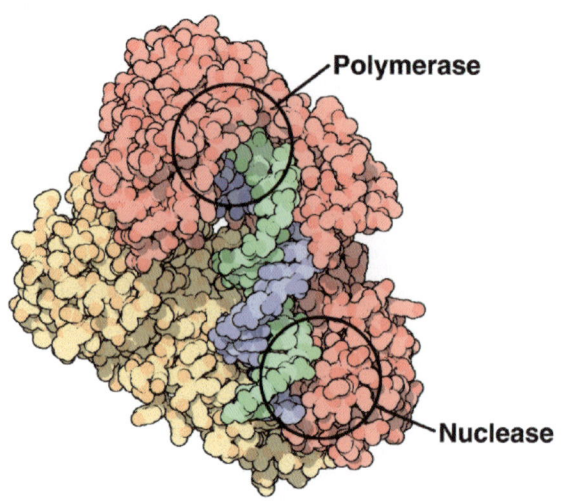

Structure of a reverse transcriptase enzyme bound to DNA (green and blue helix), showing the domains that have polymerase activity or nuclease activity.

As discussed, proteins can be purified from cells or tissues that express them. Once expressed and purified, there are many different methods available for detecting, analyzing, and quantitating them. All of these methods are used frequently in biotechnology. A significant area of research is related to discovering even more effective methods for expression, purification, and the study of proteins.

Summary

The analysis of proteins can involve utilizing many different methods, to determine the identity, quantity, or function of a particular purified protein. Currently, the most commonly used techniques include polyacrylamide gel electrophoresis, mass spectrometry, X-ray crystallography, nuclear magnetic resonance, Western blotting, and enzyme linked immunosorbent assays (ELISA). All of these methods are used in modern biotechnology.

Concept Reinforcement

1. Name one method for determining the size, structure, or identity of a protein and how this method works.

2. How are antibodies utilized to detect and analyze proteins?

3. What aspect of a protein is being detected in a functional assay?

Unit Three

Section 3.1 – Restriction Enzymes

Section Objective

- Explain restriction enzymes

What Are Restriction Enzymes?

Biotechnology relies heavily on the manipulation of DNA. Often, DNA from one organism is used in a different organism. This process of moving DNA between organisms uses unique enzymes called restriction enzymes or restriction endonucleases. The 1978 Nobel Prize in Medicine was awarded to Daniel Nathans, Werner Arber, and Hamilton Smith for the discovery of restriction endonucleases. This highlights the importance not only of their discovery, but also of their applications and key role in biotechnology.

As discussed, DNA is the repository of genetic information in the cell. This information, primarily for the production of proteins, is encoded in the specific sequence of nucleotide bases in the DNA strands. **Restriction enzymes** or **restriction endonucleases** can cut double-stranded DNA at very specific sites. They recognize a specific sequence in the DNA and make two incisions, one in each DNA strand, in the sugar-phosphate backbone of the double helix. Restriction enzymes are expressed and purified from numerous strains of bacteria that use them as a defensive mechanism to protect themselves from invasion (infection) by bacterial viruses (**bacteriophages**).

The restriction enzyme cuts the viral DNA, inactivating the virus so that it cannot replicate and damage the bacteria. This protection mechanism is called the **restriction modification system**. The bacteria protect their own DNA by modifying it such that their own restriction enzymes cannot cut it. The modification is the addition of a **methyl** group on some of the DNA bases in the recognition sequence for the restriction enzyme. Thus, for each specific DNA recognition sequence, there is a corresponding restriction enzyme/ methylation enzyme pair to internally protect the bacteria.

There are three types of restriction modification systems: Type I, Type II, and Type III. The Type I system is the most complex, and involves three proteins that work together in a group or complex to recognize the sequence, then either cut or methylate the DNA. The Type II system is the most common and also the simplest system. It involves two separate proteins, one of which is the restriction enzyme and one is the methyltransferase. They both recognize the same sequence in the DNA. The Type III system has both restriction enzyme and methyltransferase proteins that form a complex for modification and cleavage, although the methyltransferase can modify the DNA independently of the restriction enzyme. Approximately one quarter of known bacteria possess a restriction modification system. Some bacteriophages have developed systems to counteract this protection mechanism, usually by modifying their own DNA to protect it or by expression of proteins that inhibit the activity of the bacterial restriction enzyme.

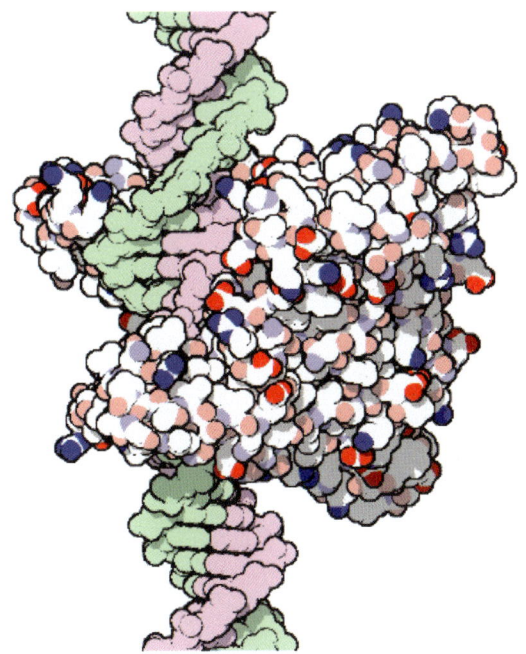

Illustration of a restriction enzyme bound to the DNA it is going to cut.
The DNA double helix is pictured in green and pink, while the enzyme is multicolored.

Restriction enzymes recognize specific DNA sequences. These recognition sites are usually 4-10 base pairs in length, with 6 being the most common. For example, the restriction enzyme "EcoRI" recognizes the DNA sequence "GATTC." Restriction enzymes are named according to the bacteria from which there were first identified. For our previous example, "EcoRI," the name is derived as follows:

E = *Escherichia* (genus of bacteria)

co = *coli* (species of bacteria)

R = RY13 (strain of bacteria)

I = first restriction enzyme identified in this strain

This naming scheme has been used on all restriction enzymes identified to date (see Table 1 for more examples). Type 2 restriction enzymes are most commonly used in research and biotechnology, as these enzymes actually cut the DNA within their recognition sequence. Other types of restriction enzymes (types 1 and 3) cut DNA from 20-100 bases down from the recognition site, which is not as useful when trying to precisely manipulate DNA fragments.

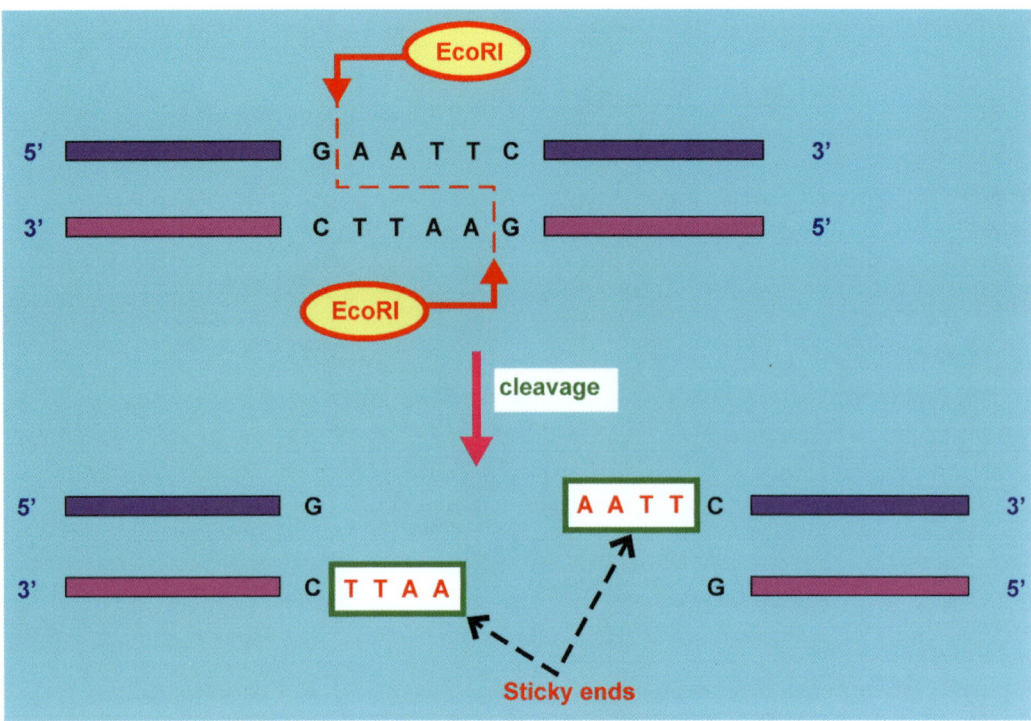

Schematic showing the recognition sequence for the restriction enzyme Eco RI, which can cut a single piece of DNA into two separate pieces that contain sticky ends.

Table 1. Examples of common restriction enzymes used in biotechnology.

Restriction Enzyme	Bacterial Source	Recognition Sequence	Resulting Product Ends
EcoRI	*Escherichia coli*	5'-GATTC 3'-CTAAG	5'-G AATTC-3' 3'-CTTAA G-5'
BamHI	*Bacillus amyloliquefaciens*	5'-GGATCC 3'-CCTAGG	5'-G GATCC-3' 3'-CCTAG G-5'
HindIII	*Haemophilus influenzae*	5'-AAGCTT 3'-TTCGAA	5'-A AGCTT-3' 3'-TTCGA A-5'
SmaI	*Serratia marcescens*	5'-CCCGGG 3'-GGGCCC	5'-CCC GGG-3' 3'-GGG CCC-5'
KpnI	*Klebsiella pneumoniae*	5'-GGTACC 3'-CCATGG	5'-GGTAC C-3' 3'-C CATGG-5'
AluI	*Arthrobacter luteus*	5'-AGCT 3'-TCGA	5'-AG CT-3' 3'-TC GA-5'
XbaI	*Xanthomonas badrii*	5'-TCTAGA 3'-AGATCT	5'-T CTAGA-3' 3'-AGATC T-5'
PstI	*Providencia struartii*	5'-CTGCAG 3'-GACGTC	5'-CTGCA G-3' 3'-G ACGTC-5'

Restriction Enzymes in Biotechnology

Once the restriction enzyme recognizes the specific sequence in the DNA, it binds and then cuts the DNA. The site of cleavage is dependent on the enzyme. Some cut the DNA directly in the middle of the recognition site (blunt end), while others cut off to the side (sticky or cohesive end). This leaves different types of DNA ends when the enzyme has completed the digestion. Depending on the recognition sequence, different enzymes might leave ends that are compatible with each other (have the same overhanging bases at each end). These ends can then be rejoined by an enzyme called **DNA ligase**. The combination of restriction enzyme digestion and rejoining by ligase has allowed for combining different pieces of DNA together and thus recombinant DNA technologies. This is the foundation of modern molecular biology and biotechnology – the ability to manipulate DNA fragments and combine them in unique and useful ways.

Many DNA fragments are at least initially joined with **plasmid** DNA for ease of purification and manipulation. Plasmids are normally found in bacteria and can be purified from them and then digested with restriction enzymes in a test tube, to allow joining of different DNA fragments. Most commercially available plasmids have been engineered (designed) to have multiple restriction enzyme digestion sites in them, usually in a specific region called the multiple cloning site or polylinker. This makes it easier to add different fragments of DNA to the plasmid, with the aid of DNA ligase.

DNA ligase

Enzyme that can attach DNA pieces together in a covalent fashion (permanent), such as after digestion by restriction enzymes.

Plasmid

Double-stranded, circular piece of DNA that is found in bacteria and yeast and is separate from the genomic or chromosomal DNA; often possess genes advantageous for different environmental conditions and can be passed from one bacteria to another.

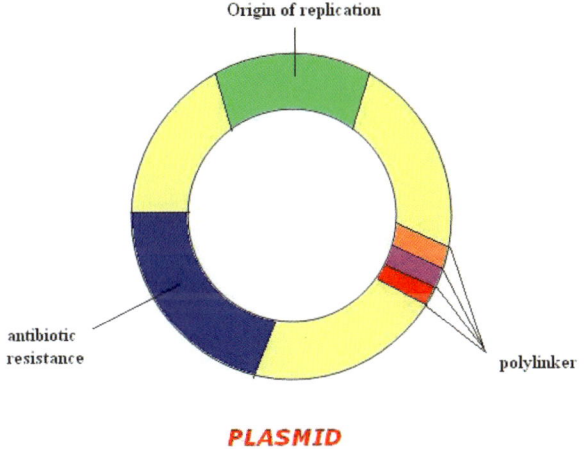

PLASMID

Schematic of a simple plasmid, containing an origin of replication, antibiotic resistance gene, and a polylinker which contains many different restriction enzyme digestion sites for ease in cloning in target DNA fragments.

Summary

Since they initially made DNA manipulation in the laboratory possible, restriction enzymes are responsible for the advent of biotechnology. They also allow foreign DNA to be introduced into other organisms, usually through the use of recombinant plasmids. Many aspects of biotechnology would not be possible without restriction enzymes.

Concept Reinforcement

1. What are the components of a restriction modification system? What is its function and what role do restriction enzymes play?

2. How do restriction enzymes cut DNA? What are the products of their action?

3. Given the following DNA sequence, develop a table that illustrates the results if that DNA sequence is cut with BamHI, SmaI, or PstI. What are the sizes of the DNA fragments that results and how many are produced by the digestion? (Note: only the top strand sequence is given. Assume that the DNA is double-stranded).

DNA sequence:

5'-ATTGCGCGTATATATCCCCGAGATATCGAGGGGATCCCTA-
ATTTGCGAGTCGGTCAGATAAATGGCGCATGATCTATGATCAG-
GACCCGGGATCCCGCGGATTAGCATGCATTGCCAGTAGCCACTT-
TATTCCCGTTAGCCAAATACTGGAACCTTATATATCCCGAGAGTT-
GACGCCTGCAGC-3'

Section 3.2 – The Polymerase Chain Reaction

Section Objective

- Explain the Polymerase Chain Reaction

What is PCR?

The **Polymerase Chain Reaction (PCR)** is one of the most widely used and powerful techniques in the field of molecular biology. Starting with very small amounts of DNA, a scientist can copy or **amplify** that DNA using PCR, resulting in millions or even billions of copies in a few hours. Without the discovery of this reaction, the field of biotechnology might not have advanced to where we know it today. It is important to note that PCR cannot copy all of the DNA in an organism at once, as dramatized in the popular the movie *Jurassic Park*. PCR is limited to copying a short section of DNA between 100 - 10,000 base pairs at a time.

The development of this technique was credited to Dr. Kary Mullis in 1983, and he was awarded the Nobel Prize in Chemistry 10 years later. During the 1970s and 1980s, scientists could isolate DNA and carry out experiments to study it, but once they ran out of their original sample, they had to go back to the sample source and purify more DNA. This was not only time consuming, but sometimes very expensive or almost impossible if the sample was extremely rare. Dr. Mullis helped solve this problem by developing a technique for copying DNA **exponentially**. Its uses today span the entire field of biotechnology, including: genetic testing, crime scene investigation or forensics, genetic medical diagnostics, mapping and sequencing the human genome, and generating genetically modified plants. In fact, most types of biological research, including biotechnology, utilize PCR.

PCR Components

Since it relies on relatively high temperatures, the PCR process depends on a thermostable DNA polymerase. This is an enzyme that strings together a new DNA strand from a template DNA strand that can withstand this heat. The DNA polymerase used in a PCR reaction comes from a bacterium called *Thermus aquaticus* (Taq), which lives in the hot springs in Yellowstone National Park. This enzyme is responsible for DNA replication in this species. Most DNA polymerases do not work very well or will degrade completely in such extreme temperatures. Taq DNA polymerase, which can withstand such high temperatures, made the PCR process possible, and more importantly, very practical.

Polymerase Chain Reaction (PCR)

Reaction in molecular biology that allows for a specific region of DNA to be copied enzymatically and exponentially to produce millions to billions of copies of that specific sequence.

Amplify

Ability to increase a signal or molecule in number, making it easier to detect.

Exponential

Logarythmic increase in the number of molecules or signal present such that it doubles with every cycle.

Hot spring, similar to the one that *Thermus aquaticus* lives in the organism from which Taq polymerase is purified from and which makes the PCR process possible

How Does PCR Work?

When performing a PCR experiment, a scientist must make sure that all of the samples being tested are treated the same and the proper controls are run simultaneously. To assemble a PCR reaction, a scientist must first prepare something called a PCR master mix. A master mix contains all of the components required for amplification of the desired DNA in addition to the template DNA to be copied. A PCR master mix contains the following components (ingredients): Taq DNA Polymerase, upstream primer, downstream primer, magnesium ions ($MgCl^2$ or $MgSO^4$), deoxyribonucleotides (dNTPs), and a buffer to provide the correct pH and salt concentration for the polymerase to be maximally active.

Primers or oligonucleotides are used to frame or identify the DNA sequence for the Taq DNA polymerase to copy. Primers are short (around 20-25 bases) single-stranded DNA pieces that are designed to specifically attach or bind to sequences flanking the region of interest to be copied. Two primers are needed since DNA is double-stranded and anti-parallel. Each primer is designed to complement the sequence 5' to 3' on opposite sides of the gene or target DNA. These PCR primers are called upstream and downstream primers or forward and reverse primers. An example of forward and reverse primers for a specific DNA target sequence are shown below:

5'-ATGGTAAAATTCATCTACCCATACTACTACTAGCAGTGGACGTAG-3'
   TACCATTT (upstream primer)
               (downstream primer)   GGACGTAG-3'
3'-TACCATTTTAAGTAGATGGGTATGATGATGATCGTCACCTGCATC-5'

To make new copies of a piece of a gene, the building blocks for the new DNA strand copies are necessary. **Deoxyribonucleotidestriphosphates**, or dNTPs, are the free nucleotides (A, T, C, G) that Taq DNA polymerase uses to assemble the new copies of the gene being amplified. dNTPs are put into the PCR master mix in excess. This ensures having enough of them for experimental conditions, making it possible to produce numerous copies of the target DNA.

The magnesium ions are included in the PCR master mix because it is very important for Taq DNA Polymerase to be highly active. The magnesium ions act as a cofactor (helper) for the polymerase. Without these magnesium ions, PCR would not be possible. In each individual PCR experiment the concentration of these magnesium ions might need to be optimized for maximal success.

The PCR Process

A polymerase chain reaction copies DNA in a series of cycles, which are repeated over and over to allow for amplification of the target DNA sequences. One cycle consists of three steps: **Denaturation**, **Annealing**, and **Extension**, or **Elongation**.

Denaturation: Raising the temperature to 94-96°C separates the strands of the double-stranded DNA into single strands by disrupting the hydrogen bonds between the complimentary base pairs.

Annealing: Lowering the temperature to 45-60°C allows the two small complementary single-stranded primers to hydrogen bond to the single-stranded template DNA, where they are complimentary; the annealed primers frame the region of interest to be copied during PCR by Taq DNA Polymerase.

Extension or Elongation: In this step, the temperature is raised to 68-72°C. This activates the Taq DNA Polymerase to extend, or elongate, the primers that were bound to the template DNA during annealing.

> **Deoxyribonucleotide triphosphates (dNTPs)**
>
> The free nucleotides building blocks used to build new DNA copies during PCR.

> **Denaturation**
>
> The first step in a PCR cycle; the hydrogen bonds between the double strand of DNA are broken by heating to 94-96°C.

> **Annealing**
>
> The second step in a PCR cycle in which the primers attach or bind to the complementary sequence on the target DNA; one primer on each strand of DNA.

> **Extension or Elongation**
>
> The third step in a PCR cycle in which the polymerase recognizes the 3' end of the bound primers and build new DNA strands by using dNTP building blocks.

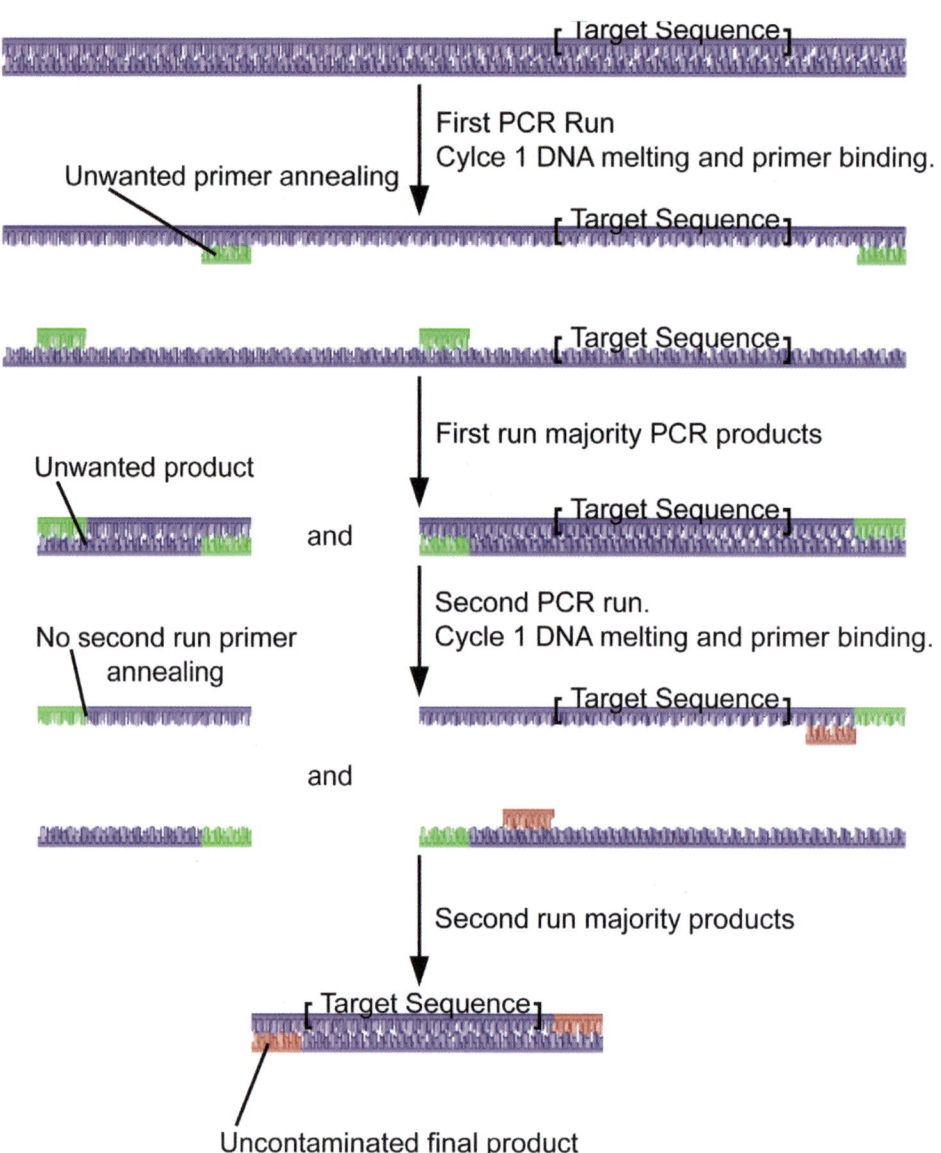

Steps in the PCR process for amplifying DNA. Visit the Dolan DNA learning center online at http://www.dnalc.org/ddnalc/resources/animations/html to see an animation of the PCR process.

Advancements in the PCR Process

Thermal cycler

A computerized heating block that can be programmed to heat and cool fast and accurately, enhancing the ease of use of the PCR process.

The PCR reaction at first took place in a series of three heated water baths, one denaturation (94-96°C), one for annealing (45-60°C), and one for extension (68-72°C). In 1986, a programmable computerized heating block was invented to perform the cycling conditions for PCR reactions; this machine is called a **thermal cycler**. Using a thermal cycler greatly simplifies use of this technique. A scientist can set up his or her reactions, program the thermal cycler, and focus on other work for a few hours while the thermal cycler and Taq DNA Polymerase produce billions of copies of their gene of interest. Thermal cyclers also added to the optimization and practicality of the PCR technique. By controlling temperatures quickly and accurately, the PCR reaction can be carried out more specifically and faster. Another factor that added to the ease of performing PCR in a thermal cycler was the development of thin walled microfuge tubes (PCR tubes) or plates. These tubes or plates provide rapid heat transfer, also making the process more efficient.

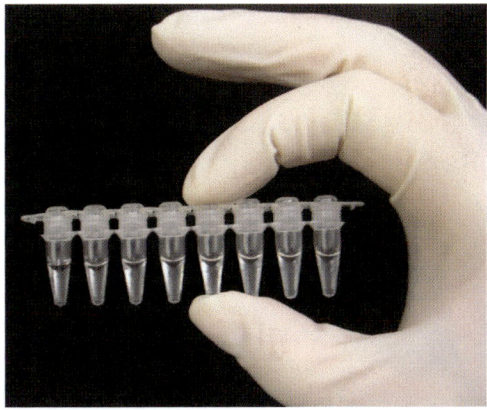

Thin walled PCR tubes containing assembled PCR reactions

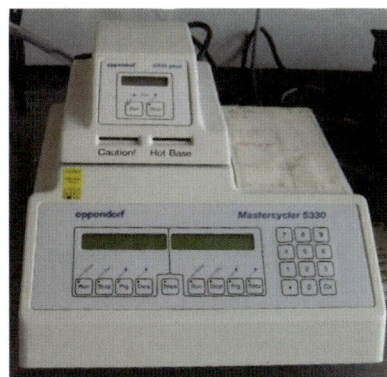

Common thermal cycler

Analysis of PCR Reactions

After a PCR reaction experiment is completed, the researcher must determine if the reaction was successful. To do this, a technique called agarose gel electrophoresis is often used, which separates DNA fragments based on size. To separate the PCR product DNA based on size, scientists take advantage of the electrical charge of DNA. DNA has a strong negative charge because of the phosphate groups that make up the sugar-phosphate backbone structure of helical DNA. An agarose gel is prepared in an electrophoresis buffer, which basically contains an ion to transfer the electrical current to and in the gel as well as to maintain the proper pH. A 1% agarose gel contains 1 gram of agarose in 100 milliliters of buffer. Agarose is a complex carbohydrate found in seaweed. It will dissolve in heated buffer but then solidify at room temperature (similar to Jell-O). Once solidified, the agarose forms a matrix, or lattice work, that the DNA can migrate through with the push of an electric current. Larger DNA fragments will have slower migration in the agarose matrix, while smaller fragments will travel farther in the gel. The larger DNA fragments remain closer to the wells in the gel where the samples were originally applied or loaded. Agarose gel electrophoresis can be used to detect most PCR products between 100-10,000 base pairs.

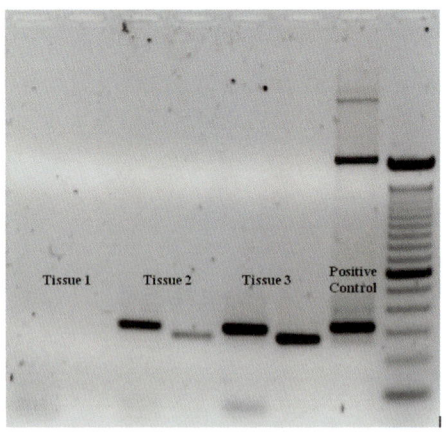

Ethidium bromide stained gel of PCR products separated by gel electrophoresis. Two sets of primers were used to amplify two different target DNA sequences from three different DNA tissue samples. No amplification was seen in sample #1 for either of the targets, while both samples #2 and #3 amplified for both targets. The positive control verified the reagents used were working properly and the last lane shows a molecular weight marker for comparison.

Applications of the PCR Process

The polymerase chain reaction is a very useful tool in biotechnology, clinical, and diagnostic laboratories. In a clinical or diagnostic setting PCR can be used to detect DNA mutations associated with various diseases or infections. In addition, PCR can be used for identifying individuals by their unique DNA signature or footprint. Since its invention, the basic PCR process has been modified and adapted to many different applications. There are numerous "twists" on the basic PCR process that continue to make this technique indispensible to biological research and biotechnology.

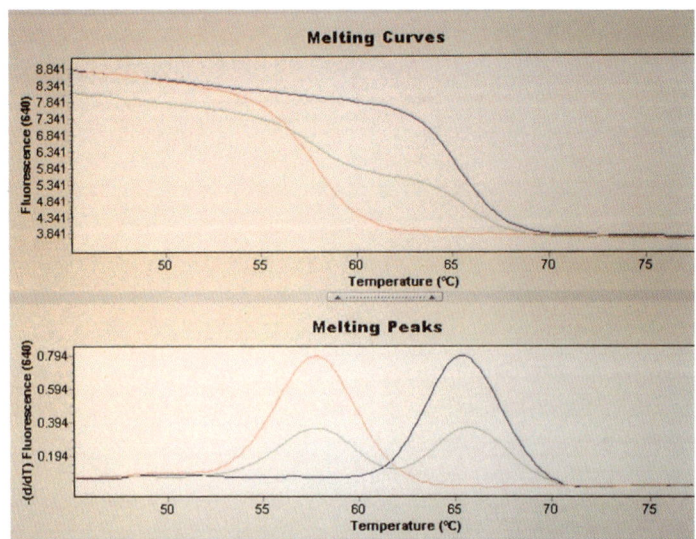

PCR used to detect mutations in the Factor V leiden gene involved in venous thrombosis.

Summary

The Polymerase Chain Reaction (PCR) has proven to be one of the most powerful techniques in molecular biology and biotechnology. Bey allowing the amplification of very small amounts of DNA, target genes can be studied and utilized by scientists in diverse applications, including genetic testing, DNA fingerprinting, recombinant protein expression, and the generation of genetically modified plants and animals.

Concept Reinforcement

1. What does "PCR" stand for and why is it an important technology?

2. What are the three main steps in a cycle of PCR and what is happening at each step?

3. If you start with 2 copies of a gene, how many copies would you have at the end of 30 cycles of PCR? (Hint: The formula for PCR is 2^n = # of copies, where n is the number of cycles.)

Section 3.3 – DNA Sequencing

Section Objective

- Describe DNA sequencing

DNA Sequencing

DNA sequencing is the ability to determine the exact order of nucleotide building blocks in a target piece or section of DNA. The sequence of DNA contains the heritable genetic information that is passed from one generation to the next and forms the basis for the development programs and blueprints in all living things. Determining the sequence of DNA is therefore very useful in basic research and in biotechnology, as well as in medical research for the diagnosis and treatment of various genetic diseases, for example. The ability to sequence DNA has greatly accelerated the process of discovery. Without sequencing, the Human Genome Project, as well as similar projects regarding other organisms, would not have been possible.

The first widely used method for DNA sequencing was developed by Allan Maxam and Walter Gilbert in 1976-1977. It was based on chemical modification of DNA and then subsequent cleavage at specific bases using defined chemicals. This method is often called "chemical sequencing," or "Maxam & Gilbert" sequencing. This method fell out of favor when newer and more convenient methods were developed, that don't involve hazardous chemicals, are less technically complex, and are easier to scale down to be more cost effective.

One of these newer methods, developed by Frederick Sanger and colleagues, is called the "chain termination" method. The key principle of the Sanger method is the use of dideoxynucleotides (ddNTPs) as DNA chain terminators during DNA synthesis. DNA is composed of repeating backbone units of deoxyribose sugar and phosphate groups, with the nucleotide bases linked to the sugar molecules. The deoxyribose sugar has a hydroxyl group at one position in the molecule that serves as the reaction site for the addition of the next nucleotide during DNA synthesis by a DNA polymerase. If this hydroxyl group is not present, as is the case of dideoxynucleotides, then a new nucleotide building block cannot be added to the growing DNA strand and thus the DNA chain is terminated. To use this method to determine the sequence of a particular piece of DNA, the template DNA is combined with a DNA primer to allow DNA polymerase to start replication, DNA polymerase, and radioactivity or fluorescently labeled nucleotide building blocks (for detection of the newly made DNA). The DNA sample is divided into four separate sequencing reactions, with each tube receiving only one of the four possible didexoynucleotide chain terminators (ddATP, ddCTP, ddGTP, and ddTTP). Incorporation of the didexoynucleotide into the newly made DNA strand stops the synthesis, resulting in various DNA fragments of varying length depending on when in the synthesis process the terminator was incorporated.

DNA sequencing

DNA sequencing is a group of methods that allow the determination of the exact sequence of nucleotide bases in a piece of DNA.

Chemical sequencing

Using specific chemicals that cleave DNA at specific sites to determine the sequence of nucleotides in the target DNA.

Chain termination sequencing

The use of didoexynucleotides to determine the specific sequence of a target DNA; uses DNA polymerase to incorporate the labeled dideoxynucleotides.

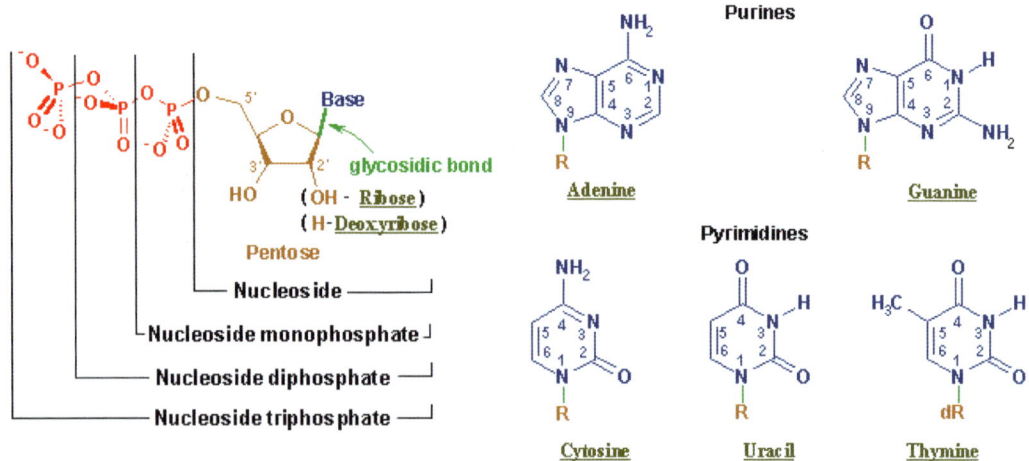

Chemical structures of nucleotide base and sugar-phosphate backbone, the building blocks of DNA. The deoxyribose is also present where the hydroxyl group is in the dideoxy nucleotides.

The newly synthesized and labeled DNA is then separated by size by gel electrophoresis so DNA fragments that are only one nucleotide different in size can be distinguished from one another. The different DNA fragments are detected either through the radioactive or the fluorescent label that has been incorporated during synthesis. For radioactive labeling, the results are dark bands on X-ray film in which each lane in the gel indicates all of the fragments that ended with the included dideoxynucleotide terminator. The sequence of the DNA is then read from the bottom of the gel to the top.

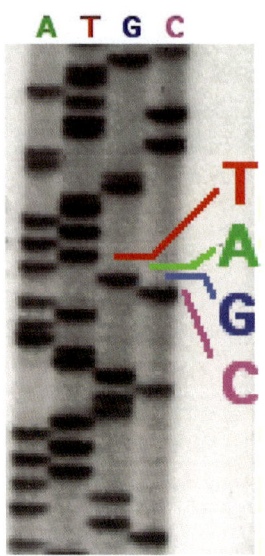

DNA sequence using radioactive dideoxynucleotide terminator sequencing. Larger DNA fragments are at the top of the picture, while smaller DNA fragments are at the bottom of the picture. Each lane indicates which different dideoxynucleotide was added to the sequencing reaction.

The chain termination method can also be used with fluorescently labeled dideoxynucleotides in the newly made DNA, which is safer and easier than radioactive detection. The resulting DNA products are still separated by size and then detected by laser excitation and the use of a fluorescent scanning machine. The results of this method appear as sequential waves on an electropherogram, with each color representing a different nucleotide. In this way, the sequence of the target DNA is read from left to right. There are commercially available reagents and kits for performing fluorescent DNA sequencing, as it has become a standard technique in molecular biology or biotechnology laboratories. Biotechnology companies offer these services to customers for a fee.

The chain termination method allows for sequencing relatively small DNA fragments, 300-1,000 base pairs in length. To sequence large DNA fragments like genomic DNA from an organism, the DNA must first be broken into smaller pieces, put into a plasmid, and then sequenced. The shorter sequences are then assembled into the longer sequence by piecing together the information.

Newer methods are being developed to meet the demand for processing many, many samples at a much-reduced cost. Many of the new high-throughput techniques use methods that combine the sequencing reactions for different templates, producing millions of sequences at once. The chain termination method, as well as many of the newer high-throughput methods, rely on the polymerase chain reaction. Some of the newer methods require not only special reagents, but also special equipment that can be quite expensive. Some of the companies producing these new systems include Roche, Applied Biosystems, Illumina, Helicos, Pyrosequencing, and Sequenom.

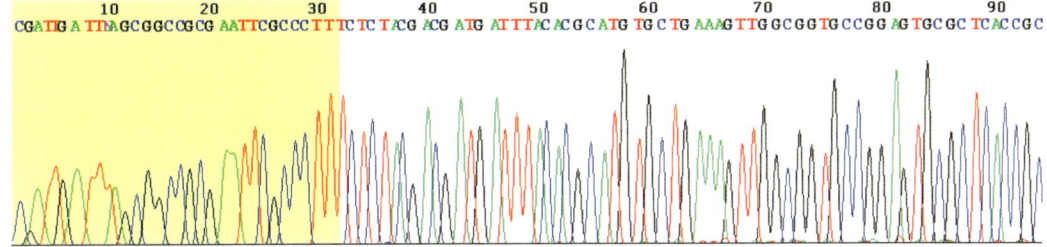

Fluorescent electropherogram from a fluorescent dye-terminator sequencing reaction. Each base is indicated by a different color, which allows for determination of the target DNA sequence. Each peak is one nucleotide different in size from the previous peak.

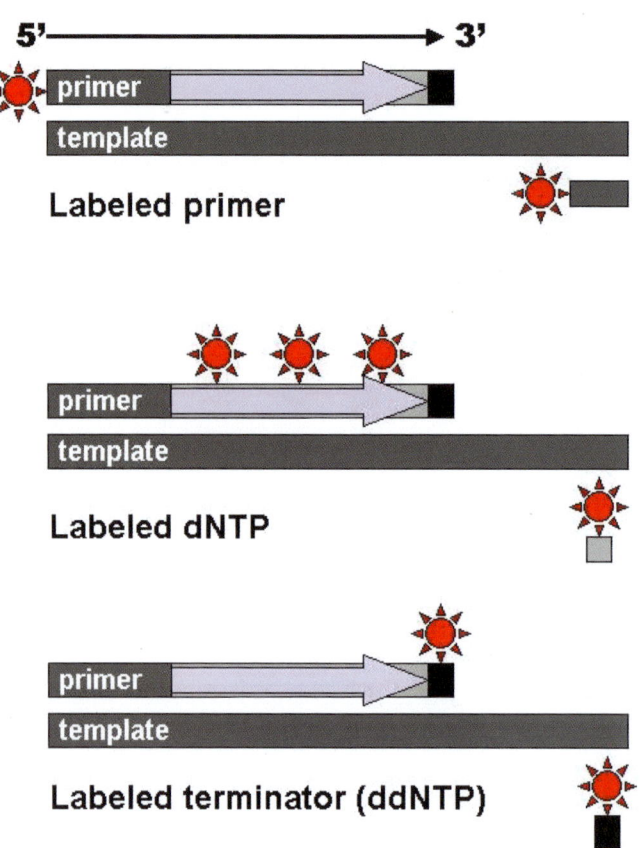

Different methods for detecting the products of terminator sequencing. The label for detecting the resulting DNA fragments may be on the primer, the nucleotide (dNTP), or the dideoxynucleotide (ddNTP).

The Human Genome Project

The **Human Genome Project** is an international scientific research collaboration. Its initial goal was to determine the sequence of DNA in the human genome, which contains approximately 25,000 genes in ~3 billion base pairs. This was accomplished over years of work supported by the United States Department of Energy, the US National Institutes of Health (NIH), and the United Kingdom's Wellcome Trust. The project began in 1990 and was initially headed by James Watson, then later by Francis Collins. A working draft of the human genome was released in 2000, and a complete one in 2003, with additional information still being published. A parallel project was conducted by a private company called Celera Genomics under the direction of Craig Venter. Most of the sequencing was performed at universities and research centers in the United States, Canada, and Great Britain, but scientists all over the world were involved.

While the initial objective was to understand the human genome, sequence efforts have been directed to several other organisms, including *E. coli* (bacteria), the fruit fly, zebra fish, yeast, and the laboratory mouse. All of these organisms have had their entire genomes sequenced. The sequence information obtained for all of these organisms aids research focusing on the function of the DNA, which genes are present and what their functions are, and how certain DNA features lead to disease.

Although the main sequencing phase of the Human Genome Project has been completed, studies of DNA variation continue in the International **HapMap project**, the goal of which is to identify patterns of single nucleotide differences (**polymorphisms**) between groups of people (**haplotypes**). This information can be very useful in medical and population migration research.

Official logo for the Human Genome Project

Summary

DNA sequencing has been critical for the emergence of biotechnology and will continue to play an important role in the future.

Concept Reinforcement

1. What is the purpose of DNA sequencing?

2. Name one method for DNA sequencing and how the technology works.

3. What is the human genome project and why is it important?

Section 3.4 – Plasmids

Section Objective

- Discuss plasmids

What is a plasmid?

Prokaryotic organisms like bacteria, and some eukaryotic organisms, such as yeast, contain DNA elements that are separate from their genomic (chromosomal) DNA. The most important extrachromosomal DNA elements that exist are called plasmids. They are typically double-stranded, circular pieces of DNA that can range in size from a few thousand base pairs (1-2 kilobase pairs or kbp) to several thousand base pairs (3-20 kbp). In bacteria, plasmids are associated with conjugation, a mechanism that enables genetic elements to be transferred from one bacterial cell to another through direct cell-to-cell contact. This process facilitates the movement of beneficial genes between bacterial cells. The benefits may include antibiotic resistance or the ability to metabolize a new nutrient that is now present in the environment. The process of conjugation was first discovered in 1946 by Joshua Lederberg and Edward Tatum, and Dr. Lederberg later coined the term plasmid in 1952.

Plasmids are capable of replicating themselves in the correct host. They contain regions of DNA recognized by DNA polymerase that allows for DNA replication to occur, similar to the process for chromosomal DNA replication. These DNA regions are called origins of replication and each plasmid will have one. Some origins of replication allow for the plasmid to be copied to a very high number in a single bacterial cell (high copy number with thousands of copies per cell), while others only allow for a limited amount of replication (low copy number with tens to hundreds of copies per cell). Some native plasmids that exist in bacteria are able to transfer between bacteria by conjugation (conjugative plasmids), while others can only be inserted by laboratory manipulation. The process of introducing research plasmids that are not able to support conjugation is an important tool in molecular biology that allows the target genetic sequence to be propagated in bacteria for a variety of reasons. In part, they are used in research because they can not be transferred to a natural bacterial population in the environment.

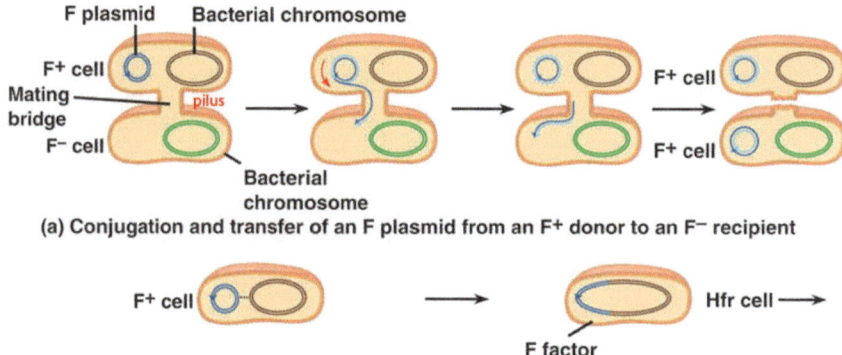

(a) Conjugation and transfer of an F plasmid from an F+ donor to an F− recipient

(b) R-plasmid carries genes for antibiotic resistance

Diagram showing the steps in bacterial conjugation, where a plasmid can be transferred from one bacteria to another through direct physical contact.

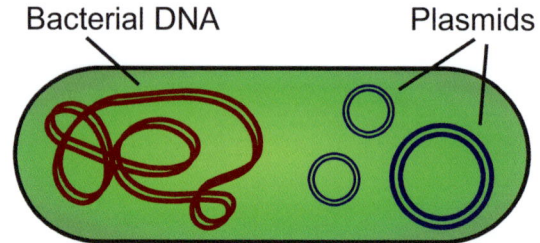

Schematic showing a typical bacterial cell with a single chromosome (bacterial DNA) and various plasmids.

Plasmids and Research

Plasmids used in research are often called vectors. There are numerous commercially available vectors that are used in research, as provided by a variety of biotechnology reagent companies. Although these vectors may have different uses, they will all have some features in common. These features include an origin of replication (ori), some type of antibiotic resistance gene (like ampicillin, tetracycline, or kanamycin), and a region that contains many closely spaced restriction enzyme cutting sites (**polyliner or multiple cloning site**) that facilitates insertion of a target DNA sequence into the plasmid. The antibiotic resistance gene allows for selection of only those bacteria that contain the plasmid, as they will be able to survive in the presence of the antibiotic, while the other bacteria without the plasmid will die. Thus the antibiotic resistance gene, and expressed protein that inactivates the antibiotic, are acting as a filter to select only the modified (plasmid containing) bacteria.

These vectors may also contain a promoter that allows for in vitro transcription, like the T7, SP6, or T3 promoters, to allow for the production of RNA corresponding to the target DNA sequence inserted into the plasmid at the polylinker site. They may contain a promoter that allows for the expression of the gene of interest as a protein in the appropriate host organism, possibly in a controlled or regulated manner, which aids in the expression

> **Polylinker (multiple cloning site)**
>
> Region of a plasmid that contains numerous restriction enzyme digestion sites to facilitate genetic manipulation.

of deleterious or toxic proteins. They may also contain some type of gene tag that facilitates target protein expression and purification, because it adds additional amino acids on to the protein that make purification easier. Another element that might be present is a reporter gene that allows for easy detection of protein expression. Such reporter genes include firefly luciferase, green fluorescent protein (GFP), or beta-galactosidase. These proteins have activities that are relatively easy to detect and generally are quantitative so that protein expression from the plasmid can be monitored and measured.

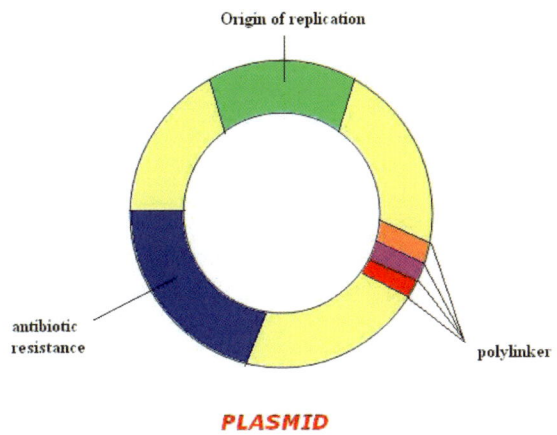

PLASMID

Typical plasmid containing an origin or replication (ori), antibiotic resistance gene, and a polylinker or multiple cloning site.

Plasmid Structures

Plasmids can exist in several different structures or conformations, all of which migrate with different mobilities when performing agarose gel electrophoresis. That is, they look like different sizes, even though they contain the same number of base pairs. Below are several descriptions of the different conformations:

Nicked open circular: one strand of the circular, double-stranded molecule is cut or nicked open; migrates slower than supercoiled.

Relaxed circular: plasmid is intact and no strands are cut, but the plasmid has had the supercoils removed by enzymatic activity, thus is relaxed; migrates slower than supercoiled.

Linear: DNA has been cut on both strands, so is no longer circular, but a linear molecule; migrates according to its size and is usually slower than supercoiled.

Supercoiled: DNA is intact with no cuts, but is twisted upon itself into a very compact and coiled structure; due to its compact structure it usually migrates the farthest as compared to the other forms.

Supercoiled denatured: DNA is supercoiled but some regions are unpaired to make it slightly less compact; usually results from excessive alkaline treatment during purification; usually migrates between supercoiled and the other forms.

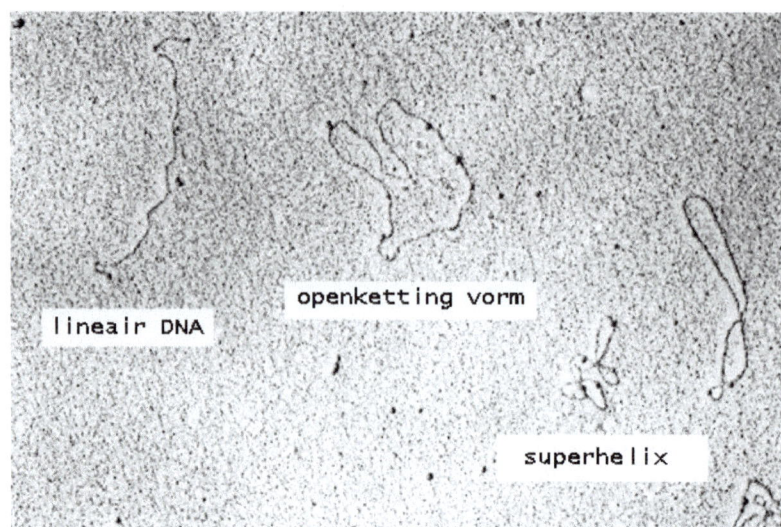

Some of the different structural forms of plasmid DNA are shown using electron microscopy: linear DNA, open circular (openketting vorm), and supercoiled (superhelix).

Plasmid Purification

Precipitate

Formation of a solid in solution during some type of chemical reaction.

Centrifugation

Use of centripetal or gravitational force to separate mixtures of molecules; more dense components migrate farther from the axis of the centrifuge and thus move to the bottom of the tube or bottle, while less dense components stay near the top; generally measured in revolutions per minute (rpm).

Plasmid DNA can be purified using a number of different techniques, but almost all are based on the alkaline lysis process. This step involves lysing (breaking open) the bacteria using a very alkaline (high pH) solution, usually sodium hydroxide that also contains a detergent. Once the bacterial cells are lysed, a solution containing high amounts of salt are added (usually potassium acetate) to **precipitate** the large debris and proteins, along with the very large chromosomal DNA. Following **centrifugation**, the plasmid DNA remains in the liquid fraction while the debris and chromosomal DNA remain in the solid pellet at the bottom of the tube. The pellet is discarded and the liquid supernatant is then further purified by precipitation with alcohol (since DNA is not soluble in alcohol) or by capture of the plasmid DNA using silica-based technologies (similar to the purification of genomic DNA or RNA). Plasmid DNA may be purified from small cultures of bacteria (miniprep; 1-10 milliliters of culture), medium-sized cultures (midipreps; 10-200 mls), or from larger cultures (maxiprep or megaprep; >200mls).

Industrial-scale plasmid purification occurs on very large scales from hundreds to thousands of liters of culture which is then purified using very large-scale chromatography systems, generally based on anion exchange instead of silica-based methods. Anion exchange chromatography takes advantage of the highly negative charge of DNA and is generally less expensive as compared to silica-based purification. The anion exchange solid support used in this type of chromatography is highly positively charged, and efficiently binds to the highly negatively charged DNA, allowing for its purification from all other cellular components. Silica (made from sand) also binds to DNA, but through weaker charge interactions.

Plasmid Analysis

The purified plasmid is then analyzed for purity and concentration using absorbance analysis. As discussed earlier, DNA will absorb light at 260 nanometers and absorbance is directly proportional to the concentration of DNA in the sample. The integrity of the plasmid DNA can be determined using agarose gel electrophoresis, either with or without first digesting the plasmid DNA with **restriction enzymes**. The use of restriction enzyme digestion is a standard method to verify the target DNA sequence has been correctly inserted into the plasmid at the correct site. The plasmid DNA may also be subjected to **DNA sequencing** to verify that the inserted target DNA is correct. This is particularly important if the target DNA is expected to express a particular protein. A mistake in the gene for the protein could result in a change in the amino acid sequence of the protein and thus its identity or activity.

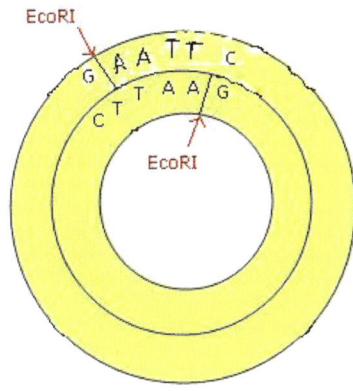

A plasmid is cut by restriction endonuclease EcoRI at restriction sites.
Plasmid DNA showing the restriction digestion site for the EcoRI restriction enzyme.
Digestion by this enzyme would convert the circular plasmid into a linear one.

Summary

Plasmids are double-stranded, circular pieces of DNA that are normally found in bacteria and yeast, but are now an important tool in molecular biology and biotechnology. Plasmids as research and industrial vectors have found widespread use in basic research and biotechnology. They provide a convenient way to clone and maintain access to target gene sequences for *in vitro* manipulation, as well as for introduction into other organisms through **transformation** or **transfection**. The target gene sequences can then be used to express recombinant proteins for use in research, industrial, or therapeutic purposes.

Restriction enzyme

Enzyme that recognizes a specific sequence of double-stranded DNA and then cuts the DNA on both strands; used in molecular biology and recombinant DNA techniques to manipulate DNA sequences.

DNA sequencing

The ability to determine the exact order of nucleotide bases in a target piece or section of DNA.

Transformation

Genetic change in a cell due to the uptake and expression of foreign DNA by that cell; there are different methods for bacteria, yeast, and plant cells.

Transfection

Introduction of genetic information, usually in the form of a plasmid, into eukaryotic cells (usually mammalian) through a number of different methods.

Concept Reinforcement

1. What is a plasmid and where do they come from?

2. How do the different forms of plasmids differ from each other?

3. How is plasmid DNA purified and analyzed to verify size, purity, and integrity?

Section 3.5 – Transformation and Transfection

Section Objective

- Compare and contrast transformation and transfection

Genetic Transformation of Bacteria

The genetic transformation of bacteria is a very useful tool for scientists in biotechnology and molecular biology. Scientists can "transform" an organism by introducing a new piece of DNA containing their gene of interest, which is usually a plasmid. Given the right conditions, the organism will replicate that piece of DNA and treat it like its own, making the product of the gene being studied. Genetic transformations can be performed in bacteria, yeast, plant cells, and animal cells. In animal cells this process is called transfection.

Bacteria do not encase their genetic material in a nucleus, which makes it easy to introduce a new gene via the introduction of a plasmid. Since DNA is basically a recipe for a protein, bacteria can take this new piece of DNA and make the protein it codes for. Bacterial transformations were first described in 1928 by an English bacteriologist, Frederick Griffith. He discovered gene transfer among virulent and non-virulent strains of *Streptococcus pneumonia* after exposure to heat. Non-virulent strains could "pick up" virulent traits by exposing them to heat-inactivated virulent bacteria. Later, in 1944, Oswald Avery, Colin MacLeod, and Maclyn McCarty demonstrated that the transformation factor was genetic in nature in *S. pneumonia* and called this uptake of DNA by bacteria "transformation."

Frederick Griffith.

> **Plasmid**
>
> Double-stranded, circular piece of DNA that is found in bacteria and yeast and is separate from the chromosomal DNA; often possess advantageous genes for different environmental conditions and can be passed from one cell to another.

The type of DNA used to perform a genetic transformation in bacteria is often called **plasmid** DNA, and as noted above, a plasmid used in research can also be called a vector. Linear DNA is usually not used because it is not stable and degrades easily. A scientist can enzymatically cut and paste new genes into a plasmid using restriction enzymes. This new gene usually codes for a protein that the researcher is studying. Examples of proteins that can be made in bacteria are human insulin, human growth hormone, and enzymes like DNase or RNase.

Conjugation

Transfer of genetic material from one bacterial cell to another by direct cell-to-cell contact; the information transferred is often beneficial to the recipient.

Transduction

Process by which bacterial DNA is moved from one bacteria to another by a virus carrier; occasionally during viral packaging, pieces of host bacterial cell DNA are accidentally packaged along with the virus DNA or RNA.

Transformation

Genetic change in a cell due to the uptake and expression of foreign DNA by that cell; there are different methods for bacteria, yeast, plant cells, and animal cells.

Competent cells

Cells that are able to take up foreign plasmid DNA from their environment; competence can be natural or induced/ artificial; artificial competence can be induced in the laboratory by treating cells with various chemicals to make them permeable to DNA.

As discussed, normal genetic transfer occurs in bacteria in nature through a process called **conjugation**. If a virus is introducing the genetic material, the process is called **transduction**. Conjugation is a natural process that allows bacteria to share plasmids with each other by direct cell-to-cell contact. It has proven very beneficial to bacteria, as it allows them to transfer traits that confer selective advantage in a particular environment, such as antibiotic resistance or the ability to metabolize a new nutrient.

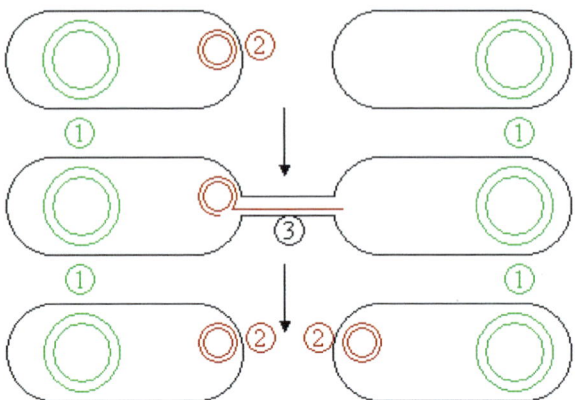

Transfer of plasmid DNA (#2) to a new bacterial cell through conjugation. The host chromosomal DNA is pictures as a large circle for comparison (#1).

In the laboratory environment, the process of introducing foreign DNA into a cell or organism is called **transformation** or genetic transformation. To perform genetic transformation in bacteria, special bacterial cells are used. They are called **competent cells** and have been treated in a way that enables them to more easily absorb DNA from their environment. Some bacteria in nature are naturally competent and able to absorb DNA in a laboratory environment (usually 1% of all species), but the use of competent cells makes genetic transformation more consistent and predictable. To make a bacterial culture competent, the cells are chemically treated with a solution of calcium chloride (CaCl2) or strontium chloride (SrCl2) during their log phase (active phase) of growth, when the cell number is doubling in ~20 minutes. They are treated in a way that neutralizes the negative charges surrounding the phospholipid bilayer of the bacteria's cell membrane. The chloride ions are permeable to the membrane and when they enter the cell through pores, water molecules from the environment also enter and the bacteria swell. The exact mechanism of the DNA uptake is unknown, but after addition of plasmid DNA the cells must be heat shocked to allow the DNA to enter. Following heat shock, the cells are incubated in growth media to allow them to survive and begin expression of the antibiotic resistance gene that is present on most plasmids. As explained in the previous section, plasmids have a few features in common, and an antibiotic resistance gene is one. They also contain origins of replication and multiple cloning sites for genetic manipulations.

Since competent bacteria are fragile, after transformation only about 1-5% of the bacteria will take up the new plasmid and survive the process. To remove or kill all of the bacteria without the plasmid, an antibiotic is added to the nutrient agar plates on which they are grown. This allows for selection of bacteria with the plasmid of interest. DNA can also be introduced into bacteria using electrical currents, in a process called electroporation.

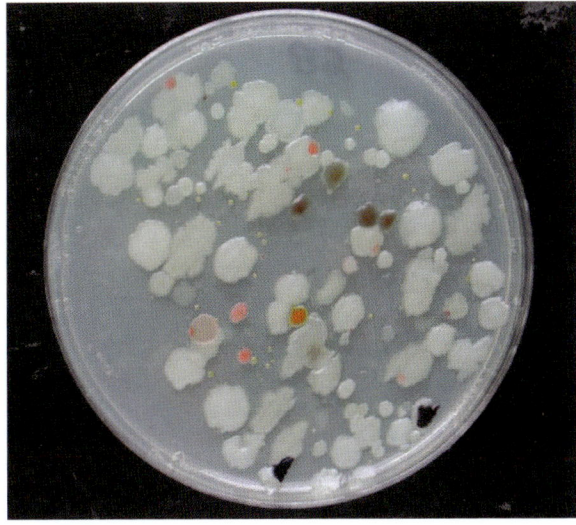

Agar plate showing bacterial growth

Genetic Transformation of Yeast

Foreign DNA in the form of a plasmid can also be introduced into yeast cells. The most common strains of yeast used in research and biotechnology are *Saccharomyces cerevisiae*, *Saccharomyces pombe*, and *Pichia pastoris*. There are several methods that are commonly used for introducing DNA into yeast cells. These include chemical methods (lithium acetate or polyethylene glycol), particle bombardment with DNA-coated gold or tungsten particles, or protoplast transformation (in which fungal spores can be converted to protoplasts by removing their protective coating and then soaking them in a DNA solution). Yeast cells can be used for protein expression of human proteins and have some advantages over bacteria, because they more closely mimic human protein expression and processing. Yeast are also used for genetic studies since they are easy to manipulate and their genomes are better understood, as compared to the human genome.

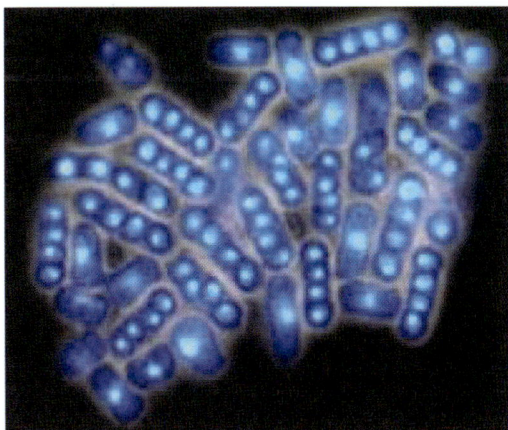

Yeast cells

Genetic Transformation of Plant Cells

A number of different methods are available to introduce DNA into plant cells. These methods are used in the production of genetically-modified plants, such as for increased herbicide resistance or nutritional value. Plasmid DNA is also used in these types of transformations. The DNA is introduced into the plant cells by electroporation, particle bombardment, viral transduction using a plant virus, or the use of a special bacterium that efficiently infects plant cells (Agrobacterium). Particle bombardment involves coating gold or tungsten particles with the DNA and then shooting them into the plant cells with a special type of machine called a gene gun. However, the most common and efficient method is the use of Agrobacterium to carry the DNA of interest into the plant cell when it is infected.

Picture highlighting plant cell transformation and the resulting plants produced with variations in their flower color.

Genetic Transfection of Animal Cells

The introduction of foreign DNA into mammalian cells is called transfection and usually involves the use of plasmid DNA. However, viruses can also be used to introduce target DNA into mammalian cells, particularly ones that are resistant to standard transfection methods. Called calcium phosphate precitation, the first transfection method was developed by F.L. Graham and A.J. van der Erb in 1973. A buffer containing phosphate ions is mixed with a calcium chloride solution containing the target DNA. When combined they form a fine precipitate of positively charged calcium and negatively charged phosphate, with the DNA bound to the surface of the precipitated particles. The cells then take up the precipitate (with DNA attached) in a process that is not completely understood. Electroporation is also used with animal cells, in which an electrical charge forces the negatively charged DNA into the cell. Cationic lipid formulations are often used to introduce plasmid DNA into animal cells. These lipids are positively charged fat molecules that coat the negatively charged DNA and allow it to fuse with the lipid-based cell membrane and allow entry into the cell.

Because plasmid DNA can not replicate in animal cells, this type of transfection is termed transient, as the introduced plasmid DNA does not remain in the cell indefinitely. It is quickly lost during repeated cell division. Plasmid DNA can be forced to integrate (insert) into the host genomic DNA (chromosomal DNA) under selective pressure, by also introducing a gene that confers resistance to a toxic compound. A small percentage of cells will insert the target DNA into their genomes. This type of transfection is then called stable, as the target DNA will remain in the cell forever, being replicated along with the host cell chromosomal DNA.

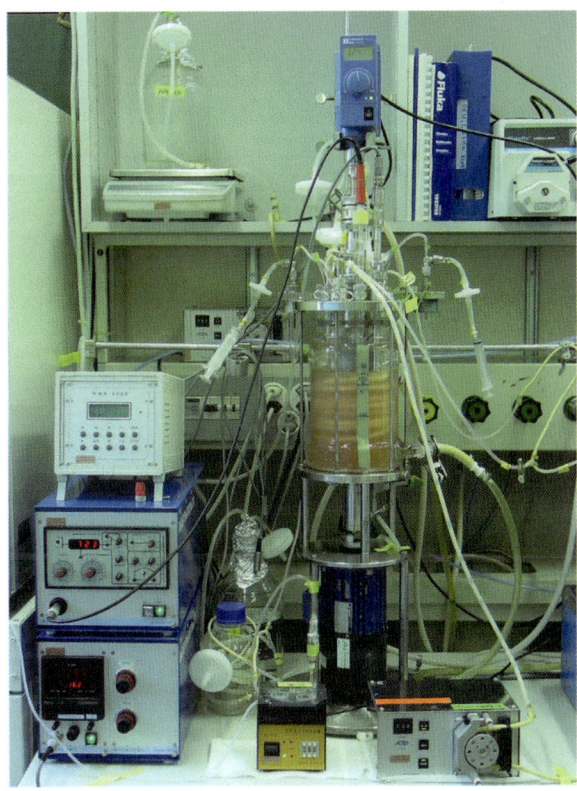

Mammalian cells stably transfected with a gene of interest to express the target protein in a larger scale bioreactor.

Summary

The introduction of foreign DNA into various types of organisms is at the heart of bio-technology and has many uses. It has been critical for the growth of biotechnology over the past several years. The expression of target genes, and thus proteins, in bacteria, yeast, plant cells, and animal cells for research or manufacturing purposes has been key. Most therapeutic drugs (proteins) are produced either in bacteria (like insulin) or in mammalian cells (like antibodies). The introduction of target genes into plant cells has enabled for the production of genetically modified organisms (GMOs) with enhanced agricultural characteristics.

Concept Reinforcement

1. How are bacteria that contain a desired plasmid separated from the bacteria that do not? Why is that important?

2. Name and explain two different methods for introducing plasmid DNA into yeast or plant cells?

3. What methods can be used to introduce foreign DNA (usually plasmid) into mammalian cells?

Section 3.6 – Recombinant DNA

Section Objective

- Define and explain recombinant DNA

What is recombinant DNA?

Recombinant DNA (rDNA) refers to a particular manipulation of DNA sequences in the laboratory that allows for the joining (or recombination) of DNA sequences that would not normally be found together. It provides a way to introduce specific target DNA sequences or fragments into different types of organisms, including bacteria, yeast, plant cells, or animal cells. It is different from genetic recombination, which occurs normally in cells and is not engineered by scientists.

The recombinant DNA technique was originally discovered and applied by Stanley Cohen and colleagues in the early 1970s, after the discovery of **restriction enzymes** by Werner Arber, Daniel Nathans, and Hamilton Smith. As discussed, these enzymes cut DNA at specific sequences, that may then be joined with other similarly cut DNA fragments, such as **plasmids**. The enzyme activity that joins DNA fragments with complimentary ends is called **DNA ligase**, and it is the key to most recombinant DNA techniques. The DNA regions that are being joined may also be generated by **PCR** (Polymerase Chain Reaction), the powerful technique discussed earlier that can amplify or copy specific target regions of DNA using a thermostable DNA polymerase and short, single-stranded DNA primers that flank the region to be copied.

The process of cloning uses recombinant DNA technology. Technically, cloning refers to creating a cell or organism derived from a parental organism that is identical to the parent. The rDNA formed is sometimes referred to as a chimera, that is, it consists of a mixture of two separate DNA sequences. Once the rDNA is produced, it can then be introduced into many types of organisms through the process of transformation or transfection.

Restriction enzyme

Enzyme that recognizes a specific sequence in double-stranded DNA and then cuts the DNA on both strands.

PCR

Polymerase chain reaction; technique that allows the amplification or copying of a specific DNA sequence into millions or billions of copies.

DNA ligase

Enzyme that can join compatible DNA ends together.

Plasmid

Double-stranded, circular piece of DNA that is found in bacteria and yeast and is separate from the genomic or chromosomal DNA.

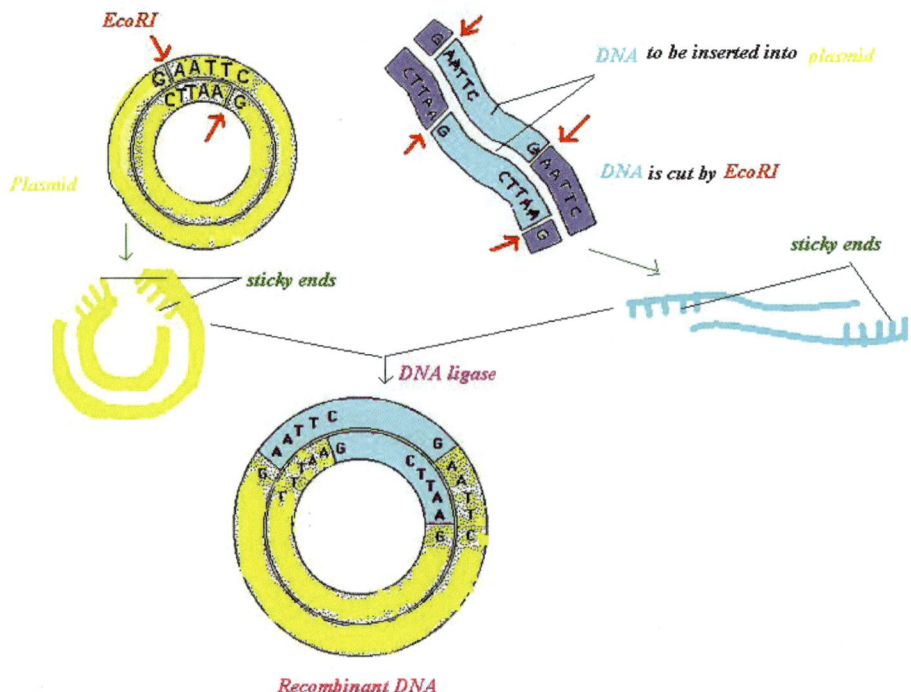

Recombinant DNA technology used to engineer a plasmid DNA to contain a new gene using restriction enzymes and DNA ligase.

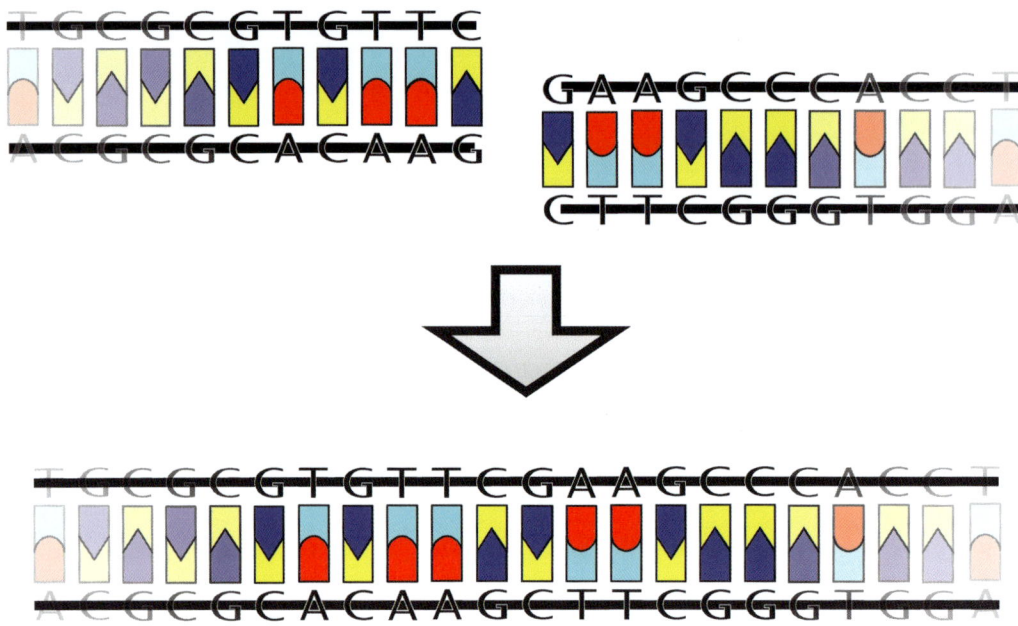

Activity of DNA ligase showing its ability to covalently link or join two fragment of DNA together into one new fragment.

The Recombinant DNA Advisory Committee

In response to public concerns regarding the safety of manipulating genetic material through the use of recombinant DNA technologies, the National Institutes of Health established the Recombinant DNA Advisory Committee (RAC) in 1974. Although RAC membership and responsibilities have changed over time, it continues to serve NIH, scientific, and public audiences concerning the scientific, ethical, and legal issues raised by rDNA, particularly in the area of clinical use. The RAC issues recommendations to the NIH director that are then conveyed through the NIH Office of Biotechnology Activities, which is responsible for the oversight of recombinant DNA research. The guidelines were first published in 1976 and are kept current based on input from both scientists and the general public. These guidelines have become a universal standard for safe scientific practice in the area of recombinant DNA and are followed voluntarily by most companies and institutions.

A major responsibility of RAC is to oversee and review human gene transfer therapy research, the process of transferring genetic material (DNA or RNA) into a person. At present, human gene transfer or therapy is still in the experimental phase. It is being studied to determine if it can be used to treat certain health problems by compensating for defective genes. Gene transfer is also being researched as a potential treatment for certain types of infectious disease, such as AIDS. Protocols that raise novel or particularly important scientific, safety, or ethical considerations are discussed by the RAC and reports are posted. The RAC is also consulted for advice concerning various advances in recombinant DNA technologies. The RAC is comprised of experts representative of a wide range of scientific and medical disciplines; it also includes ethicists, as well as members of patient and other public advocacy groups.

Applications of Recombinant DNA

There are many applications of recombinant DNA in biotechnology. Its use will continue to increase in the future as more genes and their potential commercial uses are identified. rDNA is an important tool in the generation of better crop plants (such as drought and heat resistance), recombinant DNA vaccines to prevent Hepatitis B and other viral or bacterial infections, the manufacture of protein therapeutics and drugs (such as blood clotting factors, growth factors, or hormones), and germ line or somatic cell gene therapy to treat/cure genetic diseases (such as sickle cell anemia or cystic fibrosis). To date, bacteria and mammalian cells have been used to develop more than 100 products for human medical therapy. In addition, rDNA techniques have allowed for the expression of numerous enzyme tools used for molecular biology research or industrial purposes, such as bioenergy production.

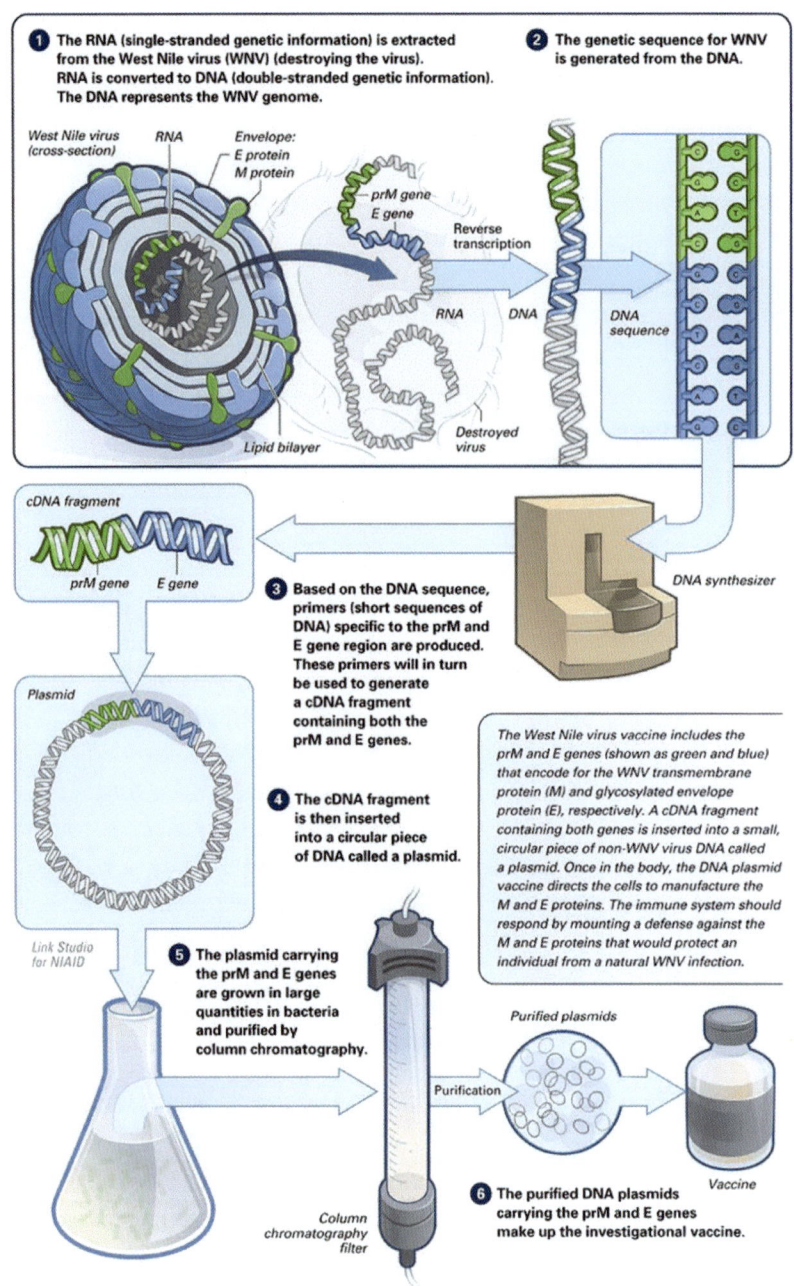

1 The RNA (single-stranded genetic information) is extracted from the West Nile virus (WNV) (destroying the virus). RNA is converted to DNA (double-stranded genetic information). The DNA represents the WNV genome.

2 The genetic sequence for WNV is generated from the DNA.

West Nile virus (cross-section)
RNA
Envelope:
— E protein
— M protein
— prM gene
— E gene
Reverse transcription
RNA
DNA
DNA sequence
Lipid bilayer
Destroyed virus

cDNA fragment
prM gene   E gene

3 Based on the DNA sequence, primers (short sequences of DNA) specific to the prM and E gene region are produced. These primers will in turn be used to generate a cDNA fragment containing both the prM and E genes.

DNA synthesizer

Plasmid

The West Nile virus vaccine includes the prM and E genes (shown as green and blue) that encode for the WNV transmembrane protein (M) and glycosylated envelope protein (E), respectively. A cDNA fragment containing both genes is inserted into a small, circular piece of non-WNV virus DNA called a plasmid. Once in the body, the DNA plasmid vaccine directs the cells to manufacture the M and E proteins. The immune system should respond by mounting a defense against the M and E proteins that would protect an individual from a natural WNV infection.

4 The cDNA fragment is then inserted into a circular piece of DNA called a plasmid.

Link Studio for NIAID

5 The plasmid carrying the prM and E genes are grown in large quantities in bacteria and purified by column chromatography.

Purified plasmids

Purification

Column chromatography filter

6 The purified DNA plasmids carrying the prM and E genes make up the investigational vaccine.

Vaccine

Schematic outlining the use of recombinant DNA techniques in vaccine development. The example here is for a recombinant DNA vaccine to West Nile Virus.

Concept Reinforcement

1. What does the term recombinant DNA mean?

2. What body governs the use of recombinant DNA in the United States and what role does it play?

3. Name one application of recombinant DNA in biotechnology.

Section 3.7 – Recombinant Proteins

Section Objective

- Discuss recombinant proteins

What are recombinant proteins?

Recombinant proteins are derived from **recombinant DNA** techniques; that is, using methods to recombine DNA fragments that are not normally found together. The DNA fragments joined usually involve a gene coding for a specific protein, combined with a **plasmid**. The gene may be obtained using the **polymerase chain reaction** or through the use of restriction enzymes from libraries of an organism or cell-specific DNA. The plasmid-gene combination can then be introduced into bacteria, yeast, or animal cells to allow that gene to be expressed into a recombinant protein. The plasmid must contain the necessary elements that allow for expression in the desired host, such as the proper promoter element for transcription. One of the most common uses of bacteria in biotechnology is for the production of recombinant proteins (and rDNA as previously discussed) for medical, industrial, or research applications. Once the recombinant proteins are expressed, they are generally purified from the host (in a semi-pure or pure state) for use.

Unique enzymes are often made that can be used in many different types of industrial processes. Enzymes made in bacteria are used to treat and modify the cloth fibers used in textile production, and in laundry detergents to boost their cleaning ability. They also help produce fuels from renewable sources of plant materials, called plant biomass. Such enzymes include cellulases, which convert cellulose, a primary component of plants, into sugars that can then be converted to ethanol by fermentation. Ethanol is used as a biofuel additive in gasoline. In the paper making industry, bacterial generated cellulose and xylanase enzymes are used to pre-treat wood and break down the lignin fibers, which saves time and energy and decreases the quantities of harsh chemicals needed to manufacture paper.

One of the most common uses of recombinant proteins is for biotherapeutics to treat a specific disease. These therapeutic proteins may be growth factors, hormones, enzymes, or monoclonal antibodies. Growth factors may be used to treat individuals who do not make sufficient quantities of the factor to support normal growth and development. This would include human growth hormone or erythropoietin (EPO) which is used to increase the generation of red blood cells in people who are anemic. One of the treatments for cystic fibrosis involves the use of an enzyme to help degrade the DNA from the cells and cellular debris accumulating in their lungs. This enzyme (DNase) is made in bacteria. People suffering with hemophilia A can be treated with the recombinant protein Factor VIII to enhance their blood clotting process. Recombinant interferons and interleukins are often given to patients who have defects in their immune systems and are unable to respond adequately to a pathogen, particularly attack by a virus.

Plasmid

Double-stranded, circular piece of DNA often found in bacteria and yeast; can be transferred between bacteria and often contain advantageous genes; used as a research tool in molecular biology and biotechnology.

Recombinant DNA

Combining (joining) DNA sequences or pieces that do not normally occur together.

Polymerase Chain Reaction

Process that allows specific DNA sequences to be copied millions of times.

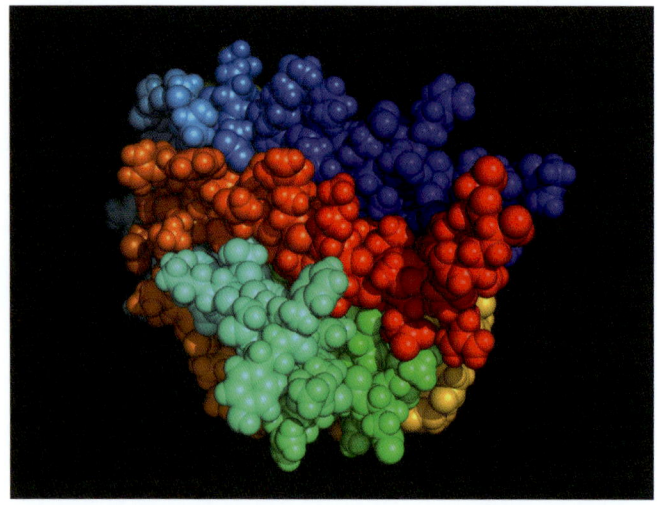

Chemical structure of erythropoietin from the crystallized recombinant protein.

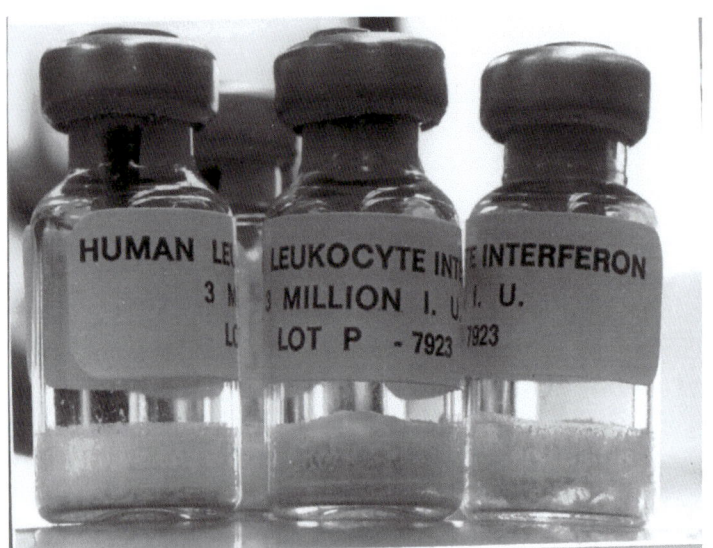

Vials of human interferon produced as a recombinant protein.

Through the production of human monoclonal antibodies, biotechnology has significantly contributed to the treatment of disease. Monoclonal antibodies are made by B lymphocytes of the immune system and recognize a specific aspect of a foreign molecule or organism (antigen). There are currently at least 25 different antibodies on the market to treat such diseases as asthma, rheumatoid arthritis, and cancer. This number will continue to increase as biotechnology and drug companies continue to invest in research and development focused on identifying novel therapeutic antibodies or proteins.

Recombinant proteins are also used in **vaccines** to illicit an immune response to a particular pathogen, such as a virus or bacteria. In the past, inactivated whole organisms were used in the development of vaccines, but now single pure viral or bacterial proteins, or mixtures of them, are used to develop vaccines. Since no whole organisms are involved, this allows for safer vaccines.

Examples of recombinant proteins used as biotherapeutics.

Biotherapeutic	Disease Target
Humira (monoclonal antibody)	Rheumatoid arthritis
Erbitux (monoclonal antibody)	Colorectal cancer
Remicade (monoclonal antibody)	Rheumatoid arthritis and Crohn's disease
Xolair (monoclonal antibody)	Asthma
Herceptin (monoclonal antibody)	Metastatic breast cancer
Erythropoietin (EPO)	Anemia
Human Growth Hormone	Short stature or growth problems
Insulin	Diabetes
Factor VIII	Hemophilia A
Factor IX	Hemophilia B
Tissue Plasminogen Activator (TPA)	Dissolving blood clots
Hepatitis B surface antigen (HBsAg)	Vaccine to Hepatitis B virus
C1 Inhibitor	Hereditary angioneurotic edema
Adenosine deaminase (ADA)	Severe Combined Immunodeficiency (SCID)
Granulocyte-Macrophage Colony Stimulating Factor (GM-CSF)	Bone marrow transplants
DNase I	Cystic fibrosis
Interferons and Interleukins	Immune system problems and viral infections

Monoclonal antibody

Antibody that recognizes a single epitope on an antigen; derived from a single B cell and its daughter cells.

Vaccine

A biological preparation that elicits an immune response to a particular disease or infectious agent or improves immunity to a particular target.

Summary

Thus, recombinant proteins play extremely important roles in biotechnology. These roles include therapeutic proteins, recombinant vaccines against viral and bacterial infections, and industrial enzymes.

Concept Reinforcement

1. What is meant by the term recombinant protein?

2. What different families (types) of recombinant proteins are often produced in biotechnology?

3. Name one specific recombinant protein and its role in biotechnology.

Section 3.8 – Antibodies

Section Objective

- Explain antibodies

Antibodies are proteins that can recognize and bind to specific shapes on other molecules, playing an important role in basic research, as well as biotechnology. Their importance as tools and therapeutics will only continue to increase in the future. This section outlines what antibodies are, what role they play in the immune system, and how they are utilized in biotechnology.

The Immune Response

The study of antibodies began in 1890, when Emil von Behring and Shibasaburo Kitasato described antibody activity against diphtheria and tetanus toxins. They proposed the theory of humoral immunity, hypothesizing that a substance in serum (blood) could react with the foreign **antigens**. Soon thereafter, Paul Ehrlich developed the side chain theory of antibody-antigen interaction, describing it as similar to a "lock and key" connection. He later received the Nobel Prize in Medicine in 1908 for his accomplishments.

This concept was further confirmed in the 1940s by Linus Pauling. In 1972 the Nobel Prize in Physiology or Medicine was jointly awarded to Gerald Edelman and Rodney Porter for deducing the structure and complete amino acid sequence of an antibody.

The vertebrate immune response is a complicated collection of internal processes that protects the organism against diseases, such as infection by bacteria and viruses, and protection from tumor (cancer) cells. It contains many layers of defense, with increasing specificity at each level. Prokaryotic organisms and simpler eukaryotic organisms also possess mechanisms to protect them from outside pathogens, such as antimicrobial peptides (small protein fragments that inhibit bacterial growth), phagocytosis (the ability of one cell to engulf another), and the complement system. However, these mechanisms are relatively simple as compared to the complex vertebrate immune system. Noting that, the complement system is also a component of the vertebrate immune system. It consists of a cascade of greater than 20 small proteins and protein fragments that interact to eventually form transmembrane channels (a membrane attack complex), that causes the targeted cells, such as invading bacteria and other pathogens, to take up water and explode.

Antibody

Special proteins expressed by cells in the immune system to recognize foreign organisms or molecules and destroy them; made up of basic structural units of heavy and light chain proteins, with constant and variable regions.

Antigen

Molecule not normally present in an organism that can stimulate an immune response; word derived from antibody generating; usually consists of proteins or carbohydrates (sugars) and not lipids or nucleic acids.

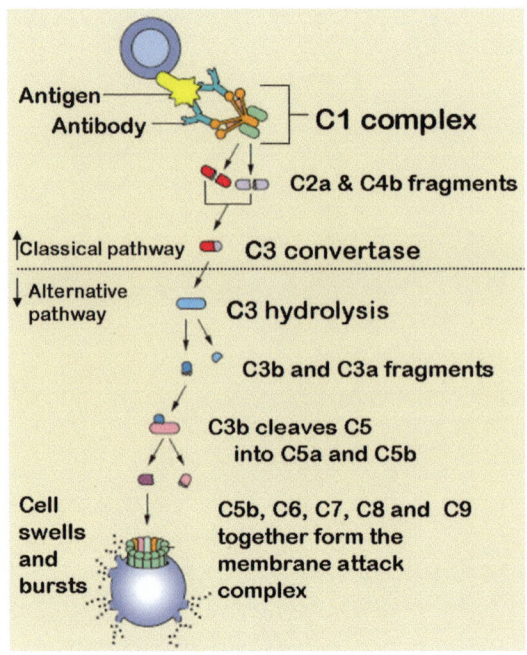

The complement cascade for lysing invading cells.

The immune system of vertebrates consists of many types of proteins, cells, organs, and tissues, which interact in an elaborate and dynamic network. The immune system adapts over time to more effectively recognize pathogens previously seen. This ability to adapt to increase effectiveness over time is why our immune system is so powerful and such a marvel of evolution. However, problems can arise when the immune system is not active enough (immunodeficiency) or hyperactive (autoimmune diseases), both of which can be critical conditions in human and animal health. Immunodeficiency can result in recurring and life-threatening infections, while autoimmunity can result in the immune system attacking normal tissues such as cartilage (rheumatoid arthritis) or the pancreas (type I diabetes).

Immune System Components

The immune system can be divided into two basic components: 1) innate (passive) immunity, and 2) adaptive (acquired or active) immunity. Passive immunity is transferred from a mother to her child through the placenta prior to birth and through bread-feeding. Immunity may be acquired naturally through exposure to an environmental pathogen, or through deliberate exposure to the foreign agent or injection with antibodies.

Characteristics of the immune system: Innate versus Adaptive.

Innate Immunity (Passive)	Adaptive Immunity (Acquired or Active)
Response is non-specific	Pathogen and antigen specific responses
Exposure leads to immediate maximal response	Lag time between exposure and maximal response
Immunity is short lived (few months)	Immunity is much longer, sometimes life-long
Cell-mediated and serum (humoral) components	Cell-mediated and serum (humoral) components
No immunological memory (no anti-bodies formed)	Exposure leads to immunological memory (antibodies are formed)
Found early in all multi-cellular life forms	Found only in jawed vertebrates

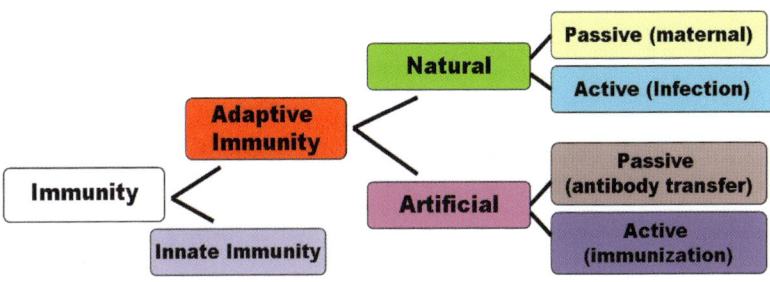

Flow chart demonstrating general features of adaptive versus innate immunity.

Physical barriers, including mechanical, chemical, and biological, reduce infection. If a pathogen does breach this first defense, the innate immune system provides an immediate, but non-specific, response to rid the organism of the pathogen. This system is found in all plants and animals. If the innate immune response is evaded, then the adaptive immune system is activated to fight the infection. This third level of defense adapts its response during an infection to improve its recognition of the foreign agent in the future, and provide a faster and stronger attack if the same pathogen is encountered again (called immunological memory). The process of acquired immunity is the basis for vaccination, and is why vaccination is so successful at preventing serious infection by dangerous pathogenic bacteria or viruses.

Both innate and adaptive immune systems rely on the ability of the body to distinguish between self and non-self molecules. Self molecules are those components of an organism's body; anything else is considered non-self (foreign) and called an antigen. The innate immune system is usually triggered when an organism recognizes specific general molecular features of microorganisms, and thus the response is generic. Through the production of chemical factors called cytokines, immune cells (neutrophils, mast cells, macrophages, basophils, eosinophils, other leukocytes, and lymphocytes) are recruited to the site of infection or irritation. This response induces inflammation at the site. The complement cascade is activated to kill the invading microorganism and to further activate other cells of the immune system involved in clearing dead cells (phagocytes). The innate immune system also activates the acquired immune system through a process called antigen presentation. Cells of the innate immune system, particularly macrophages, engulf and digest pathogens and then present on their cell surfaces molecular pieces of the pathogen (called **antigens**). These antigens are then recognized by cells of the acquired immune system.

Various types of white blood cells.

Acquired or adaptive immunity evolved early in vertebrates and allows for a stronger immune response, as well as immunological memory, so that pathogens are remembered by a signature antigen. The response is antigen-specific and requires recognition of specific non-self antigens during a process of antigen presentation by cells of the innate immune response, such as macrophages. The cells of the acquired immune system consist of a special type of leukocyte called lymphocytes. B cells and T cells are the major lymphocytes, involved with antibody production and involved in a cell-mediated immune response, respectively. Cell-mediated immunity involves the ability of special cells to actually kill other cells or signal killer cells to do so. Natural killer (NK) cells are the third type of lymphocyte and are able to recognize infected or tumor cells and kill them.

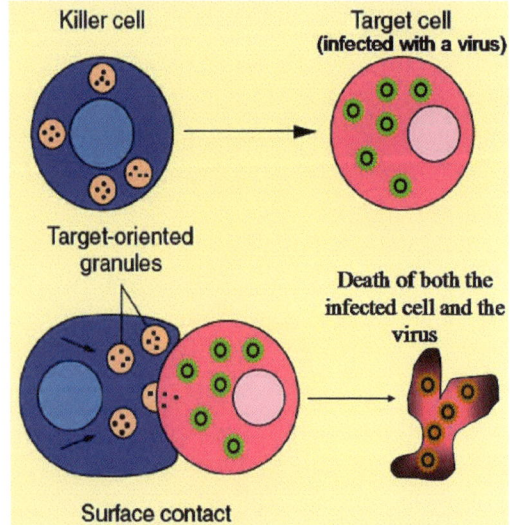

Interaction of a natural killer cell with an infected target cell, causing its death.

B cells recognize antigens on the surface of pathogens without any need for further antigen processing or presentation by the innate immune system, such as is required for T cell mediated immunity. Each lineage of B cell expresses a unique and different antibody, so that a complete set of B cell antigen receptors represent all of the antibodies that the body can produce. Once an antigen is recognized by both a B cell and a helper T cell, the B cell becomes activated to divide (reproduce itself), and these daughter cells (called plasma cells) secrete millions of copies of the specific antibody that recognizes the specific antigen.

Antibody Structure

Antibodies (also called immunoglobulins) are **glycoproteins** made up of two heavy and two light chain proteins connected by **disulfide bonds**. They have a molecular weight of about 150,000 and tend to have a Y shape. Within these chains are constant and variable regions. The variable region contains a section that is extremely variable (hypervariable) that is responsible for recognizing and binding to a particular antigen, similar to a lock and key. This huge diversity of antibodies allows the immune system to recognize an equally diverse set of antigens. Antibodies all differ in the variable and hypervariable regions, but have the same constant regions within an isotype, or class, of heavy chain expressed. There are five different isotypes or classes of antibodies, and they perform different functions at different locations within the body (see chart, page 222).

They circulate in the blood and lymph systems. When bound to a pathogen, they mark it for destruction and clearance from the body, or prevent it from entering a target cell. The lymph or lymphatic system is similar to the circulatory system, but instead of carrying blood it transfers a clear fluid called lymph. It is responsible for removing extra fluid from tissues and acts to carry fats and components of the immune system throughout the body. In addition to the five different types of heavy chains expressed, there are also two different light chains expressed (kappa and lambda). This also adds to the incredible diversity of antibodies that are generated during development and which are available for pathogen selection. It has been estimated that humans generated approximately 10 billion different antibodies, each capable of binding to a specific **epitope** of an antigen.

This extreme antibody diversity is derived from a limited number of heavy and light chain genes, which during antibody generation undergo complex genetic recombination.

Glycoprotein

Protein with carbohydrates (sugars) attached; often found in the cell membrane or outside the cells as secreted products.

Disulfide bond

Attachment between the sulfide (thiol) groups of two cysteine amino acids in a protein or between different proteins; stable in an oxidizing environment, but unstable in a reducing environment, where they become unlinked.

Epitope

Specific feature or part of a molecule that is recognized in an antigen by the immune system; an antigen can have multiple epitopes, since it may be large and have many parts that the immune system recognizes are foreign.

Different isotypes of mammalian antibodies.

Isotype Name (class)	Number of Subtypes	Description
IgA	2	Found in mucosal areas, such as the GI tract, respiratory tract, and urogenital tract to prevent colonization by pathogens; also found in saliva, tears, and breast milk.
IgD	1	Function less well defined as compared to other isotypes; seems to be involved as an antigen receptor on B cells that have not been exposed to antigen.
IgE	1	Binds to allergens and triggers histamine release from mast cells and basophils in an allergic reaction; also seem to provide some protection against parasitic worms.
IgG	4	The major antibody of the immune system against invading pathogens; only antibody able to cross the placenta to give passive immunity to the fetus.
IgM	1	Expressed on the surface of B cells and in a secreted form as well; has a very high avidity (attraction) for antigen; eliminates pathogens in the early stages of humoral immunity before there is sufficient IgG.

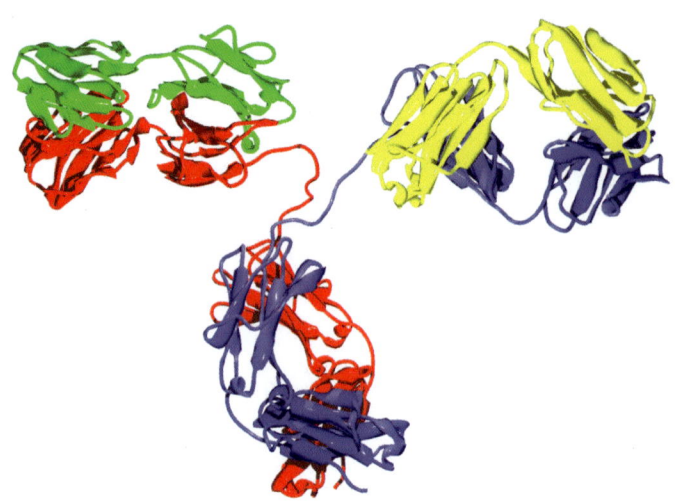

Schematic of an antibody showing the two heavy chains (red and blue) and the two light chains (green and yellow).

In addition to mounting an immune response to foreign pathogens, the immune system response also provides some protection from tumor or cancer cells. Tumor cells often express different proteins on their surface that appear as foreign to the immune system which can therefore recognize and attack them using killer T cells. Unfortunately however, many tumor cells have evolved mechanisms to evade this response, as have many pathogens.

Antibody Isolation and Purification

Antibodies can act as very specific tools to identify a specific antigen, therefore it is useful to identify and purify them for use in research or therapeutics. Specific antibodies are often produced in animals (like mice, rats, rabbits, sheep, goats, or horses) by injecting them with the specific antigen (purified proteins or inactivated pathogen). The blood isolated from these animals contains **polyclonal** antibodies for the antigen. That is, multiple antibodies that bind to the same antigen (molecule), but different epitopes (sites or features) on that antigen. Antibodies can also be produced in chickens and isolated from their egg yolks. The antibodies are often purified using proteins from bacteria known to bind specifically to antibodies, called Protein A or Protein G, that are linked to some type of solid support for chromatography.

To isolate or purify a single antibody to a single epitope on an antigen, a **monoclonal** antibody must be produced. This involves isolating the antibody-secreting B cells from an animal and then fusing these cells with a cancer cell line that allows the B cell to grow indefinitely in vitro, while secreting antibody in their growth medium environment. Single hybridoma (fusion) cells are isolated by diluting the pool until individual cells are separated and can be screened for the desired antibody. The process of generating monoclonal antibodies by making hybridomas won Niels Jerne, Georges Kohler, and Cesar Milstein the Nobel Prize for Medicine in 1984.

Antibody Testing and Use in Biotechnology

Antibodies are then tested to verify that they identify the correct epitope or antigen. This testing may involve Western Blot analysis or ELISA type assays, which require that an antibody recognize a specific antigen to be successful. In each case, a detectable signal will not be generated unless the antibody binds to the target antigen.

In research, antibodies have many different applications. They are often the only tools available to identify specific proteins and other biomolecules. Besides Western Blot analysis and ELISA assays, they are also commonly used as tags to identify specific intracellular and extracellular proteins in a cell or organism to gather information about the location and movement of the target protein. This process is called immunocytochemistry or immunohistochemistry, depending on whether the sample being analyzed is composed of cells or tissue, respectively. The antibody is usually detected by a colorimetric or fluorescent method, using a special microscope on cells or tissue fixed on glass slides. Antibodies may also be used for identification and purification of antigens from complex solutions or cell lysates.

Polyclonal

Pool of antibodies that recognize a single antigen, but multiple epitopes; derived from many different B cells.

Monoclonal

Antibody that is specific for a single epitope; derived from a single B cell and its daughters.

This process is called immunoprecipitation and not only allows for the target protein to be captured and analyzed, but any other proteins or molecules that might bind to it. This procedure is particularly useful in identifying proteins that bind to other proteins to form complexes that have important cellular functions.

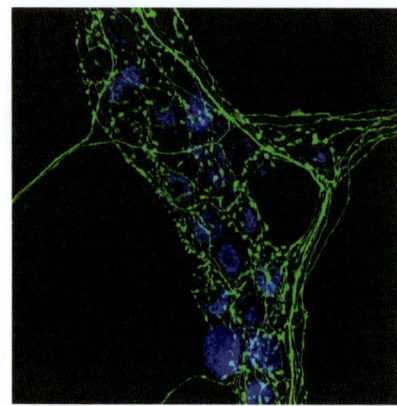

Immunocytochemistry of mouse neural cells stained with an antibody specific to a neural protein (tyrosine hydroxylase) which is shown in green.

Antibodies are becoming more common as therapeutics and are the most common biotechnology drugs. Therapeutic antibodies are monoclonal antibodies that target a specific protein within the body and, upon injection into a human, bind to that target protein and induce a physiological response. The first antibodies themselves were often seen as foreign to the human immune system, but now humanized or human antibodies are used for disease treatment. A very new strategy using monoclonal antibodies involves linking a drug-activating enzyme to the antibody that recognizes a cancer cell. Use of ADEPT (antibody-directed enzyme pro-drug therapy) has been limited to date but holds great promise for the future.

Summary

The immune response if a complex system involving cells, antibodies, and signaling factors. The immune system, and antibodies in particular, are incredibly important tools or products used in biotechnology today, and will continue to play an important role in the future. They have allowed scientists to study specific cellular functions or have been used as therapeutics to treat cancer, arthritis, and asthma to name just a few.

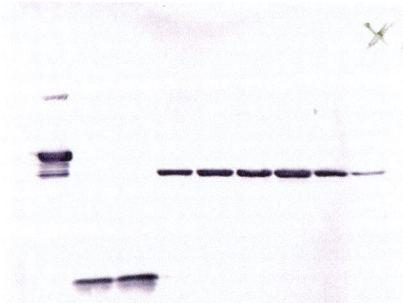

Western blot showing specific target proteins recognized by a specific antibody.

Concept Reinforcement

1. What is the purpose of the immune system? What organisms have the most advanced immune system?

2. What is the difference between an antigen and an epitope?

3. How many classes of antibodies are there and what are their names? How are they different?

Section 3.9 – Bioremediation

Section Objective

- Discuss bioremediation

What is bioremediation?

Bioremediation is any process that uses microorganisms (mainly bacteria), fungi, green plants, or their enzymes to return the environment to its natural state by removing toxic chemicals, oil, gasoline, heavy metals, or radioactivity. It takes advantage of biological organisms and their normal enzymatic or metabolic processes to remove contaminants from water or soil. When green plants are used to clean up the environment it is called phytoremediation, and when fungi are used, mycoremediation. Bioremediation is nature's way to a cleaner environment!

> **Bioremediation**
> The use of microorganisms (mainly bacteria), fungi, green plants, or their enzymes to return the environment to its natural state by removing toxic chemicals, oil, gasoline, heavy metals, or radioactivity.

Bioremediation has occurred for centuries by the action of bacteria, fungi, and plants normally found in the environment. The use of plants to extract excess salt from agricultural land has a long tradition in parts of the world, a process called phytoextraction. During the 1960s, the directed use of microbes to specifically clean up the environment was first investigated by George Robinson, a petroleum engineer in California. Bacteria can be used to degrade chlorinated hydrocarbons from soil contaminated with dangerous chemicals or to clean up oil spills. Nitrate and/or sulfate fertilizers are added to facilitate the decomposition of crude oil by indigenous or exogenous bacteria at the site of the spill. In general, the byproducts of bioremediation are water and harmless gases such as carbon dioxide.

Although many types of contaminants can be removed or at least reduced by bioremediation, not all contaminants can be easily treated with microorganisms. Heavy metals, such as cadmium and lead, are not easily absorbed or captured by bacteria. In this case, phytoremediation can be quite useful, because natural or transgenic plants are able to bioaccumulate (store) these heavy metals in their plant structures, which can then be removed for disposal or recycling.

As indicated above, mycoremediation is a form of bioremediation that uses fungi to remove contaminants, usually from the soil. The significant challenge of mycoremediation is to determine the right fungal species for a specific pollutant. Certain fungal strains have been reported to be capable of degrading nerve gas and sarin, both highly toxic biological weapons. In an experiment to test the ability of fungi to remove polycyclic aromatic hydrocarbons, diesel oil was placed on a plot of soil that contained oyster mushrooms. A control plot contained traditional bacteria. After four weeks, more than 95% of many of the contaminating hydrocarbons from the diesel oil had been degraded to non-toxic byproducts in the plot containing bacteria and the mushrooms. Wood-degrading fungi have been shown to be particularly effective at breaking down aromatic pollutants from petroleum products and chlorinated compounds from pesticides.

Bioremediation has many advantages over conventional environmental cleanup. It can be used in areas where physical removal would be very difficult. It can be less expensive and more efficient, and also does not require the contaminated area be removed, minimizing the disturbance of the site and any associated potential spread of contamination. Often the main disadvantage of bioremediation is the time it takes for the organisms to decontaminate a site. This is highly dependent on the type of contamination, the scope (severity of the spill), and whether the cleanup needs to occur above ground or underground.

The United States Geological Survey (USGS) and the Environmental Protection Agency (EPA) have played important roles in researching and implementing bioremediation. After passage of the Superfund legislation for environmental cleanup of toxic waste, Congress directed the USGS to conduct a program to provide critical information about how bioremediation could be utilized. They systematically investigated many sites throughout the United States. One of the principle findings of their studies was that microbes in shallow water aquifers affect the fate and transport of virtually all kinds of toxic substances that contaminate our environment.

Examples of Bioremediation

One example of the application of bioremediation is the Hanahan Bioremediation Project, conducted in Hanahan, South Carolina. Hanahan is a suburb of Charleston located very close to a military base at which a massive fuel leak occurred in 1975. The original spill was physically contained, but this action could not prevent fuel from leaking into the sandy soil and reaching the underlying water table. Unfortunately, the ground water began leaching toxic chemicals from the spill, carrying them toward a nearby residential area, reaching it in 1985. How could the contaminated ground water be kept from further seepage toward the residential area? The key was to use bioremediation, since earlier work by the USGS had shown that microorganisms normally present in the soil can transform harmful chemicals into harmless carbon dioxide, particularly if they were given nutrients to stimulate their growth and metabolism. In 1992, nutrients were delivered to contaminated soils and contaminated ground water. By the end of 1993, the contamination in the residential area had been reduced by 75% and even more at the sites where the nutrients were directly placed. Bioremediation had been successful!

Other examples of natural bioremediation have also been documented by the USGS, including treatment of a crude oil spill in Minnesota, sewage leakage in Cape Cod, chlorinated solvents in New Jersey, pesticides in San Francisco Bay, agricultural chemicals in the Midwest, and creosote and chlorinated phenols used as wood preservatives in Florida. Fortunately for us, microorganisms can adapt to extremely harsh chemical conditions.

In some instances the normal microbes present in the contaminated area are used, with the addition of nutrients and sometime air, while in other circumstances special microbes might also be added if needed to more effectively decontaminate the site.

Organisms such as bacteria can be used to clean up oil spills through bioremediation.

Phytoremediation of a former gas station in Denmark.

Mycoremediation: slime mold growing on wet office paper waste.

Summary

The future of bioremediation looks quite bright, although it is important to recognize that much work still needs to be done. The ability of biotechnology to engineer organisms that can effectively and efficiently remove various types of contaminants from the environment is a very exciting field of study and one that biotechnology companies, universities, the government, and the general public are quite interested in.

Concept Reinforcement

1. What is bioremediation and how it is used today?

2. What are phytoremediation or mycoremediation and how are they used?

3. Give one specific example of bioremediation.

Section 3.10 – Bioenergy

Section Objective

- Describe bioenergy

What is bioenergy?

Bioenergy is renewable energy derived from a biological source. It is often used in reference to **biofuels**, which are derived from various biological sources, such as biomass from plants or animals. The energy contained in **biomass** is basically the energy from sunlight stored in a chemical form. Biotechnology is now trying to harness it to replace our non-renewable sources of energy (fossil fuels in particular). Biomass sources include wood and wood waste, straw, manure, miscanthus, sorghum, cassava, jatropha, rapeseed, sugar cane, sugar beets, corn and corn byproducts (like corn stover), soy and soy products such as oil, grasses, palm oil, molasses, and many others. Basically, any source of carbon can be considered, but plant material is most often utilized for this purpose. Some of the plant sources listed above may not be familiar to you. Miscanthus is a perennial grass that grows well in parts of Asia and Africa. Sorghum is a grass often raised for its grain in warmer climates. Cassava is a woody shrub found in South America, that is often harvested for its edible roots, while jatropha is a family of succulent shrubs and trees native to the tropics. The seeds of jatropha plants may be useful in biofuel production because they can contain up to 40% oil.

> **Bioenergy**
> Renewable energy derived from biological sources of fuel.
>
> **Biofuel**
> Fuels derived from biological sources (biomass); include ethanol, biodiesel, plant oils, wood pellets, etc.
>
> **Biomass**
> Organic material derived from living organisms, including plants, animals, and their byproducts.

Corn ethanol plant in Iowa.

The need for biofuels, as well as other alternative sources of energy, has increased dramatically in the past several years due to the increased cost and decreased availability of fossil fuels such as gasoline, diesel, and jet fuel. It is estimated that the crude oil reserves around the world will be depleted in approximately 40 years. In addition, as the reality of climate change is recognized, the need to drastically reduce carbon emissions into the environment is being addressed. Since it uses renewable biological plant matter as the energy source, utilizing bioenery contributes to the solution to these challenges. Energy derived from plants has the huge advantage of generally being carbon neutral. The carbon released when the biofuel is used is reabsorbed into new plants as they grow, and no excess carbon dioxide is released in this process.

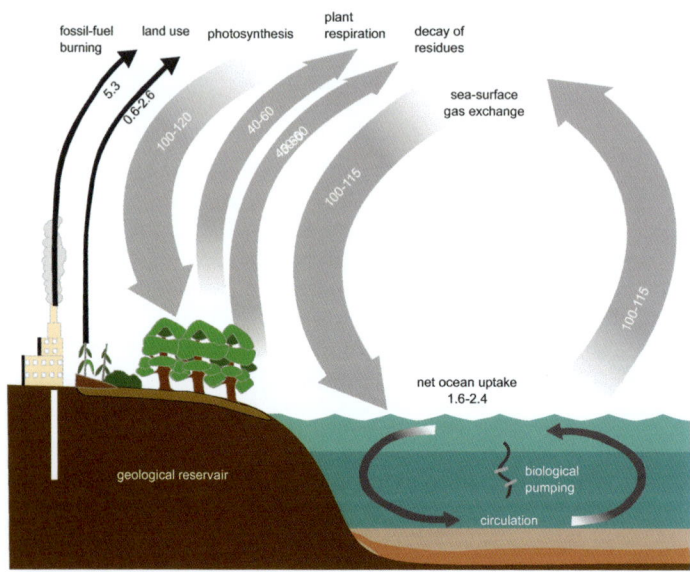

Simple representation of the carbon cycle on earth.

Brazil uses biomass from sugar cane fibrous waste (bagasse) to not only produce ethanol, but also electricity. The production of sugar cane in Brazil takes full advantage of every ounce of stored energy in the sugar cane. After processing the cane for sugar, the bagasse is burned at the mill to provide heat for distillation of the ethanol produced from the fermented sugars, and for electricity to run the processing machinery. These Brazilian plants are energy self-sufficient and, moreover, any excess electricity generated is supplied to local utilities. Another way to produce electricity from biomass is called co-firing, in which biomass is added to coal to generate energy. The biomass generally represents 1-15% of the input at the coal plant. Using biomass with coal aids in reducing the damaging sulfur and nitrogen emissions and actually increases the efficiency of the energy generating boilers.

Anaerobic digestion of biomass, such as agricultural and forestry wastes, can be used to produce biogas or gasification to produce syngas for direct combustion. Methane is often produced as a natural byproduct of anaerobic bacterial digestion of organic garbage and is being trapped at some landfills for use as an energy source.

Types of Bioenergy or Biofuels

As mentioned, the most common sources of biomass to create biofuels are plants. There are two very common strategies for generating bioenergy from biofuels. The first is either grow crops high in sugar or starch content, which can then be fermented by bacteria or yeast to product ethanol, propanol, butanol, and methanol. The second is to grow crops that produce oil that, after being heated to reduce viscosity, can be burned directly in a diesel engine or chemically converted to biodiesel through a process called transesterification. Biodiesel is compatible with most diesel engines manufactured after 1994, and most diesel cars and trucks in Europe use as least part biodiesel fuel. A 5% biodiesel blend is quite common in many European countries. Since 80% of the trucks and buses in the US run on diesel fuel, the production of biodiesel in the United States is expected to increase dramatically in the near future.

Other enzymes may be involved to release fermentable sugars from the plant biomass, since they are often bound up by insoluble plant compounds. Cellulose can be converted to fermentable 6-carbon sugars, but must first be released from a cell wall scaffold composed of hemicelluloses, pectin, and lignin. This is often done using harsh heat, chemicals, or special enzymes called cellulases. These enzymes can be obtained through biotechnology, and research efforts are trying to increase the efficiency of these fermentation and enzymatic processes. The goal is to reduce overall costs so bioenergy can be cost competitive with fossil fuels. A new process called Simultaneous Saccharification and Fermentation has greatly improved ethanol production efficiency. In this process, cellulose enzymes and fermentation microorganisms are combined in a single reaction mixture to produce ethanol in one step.

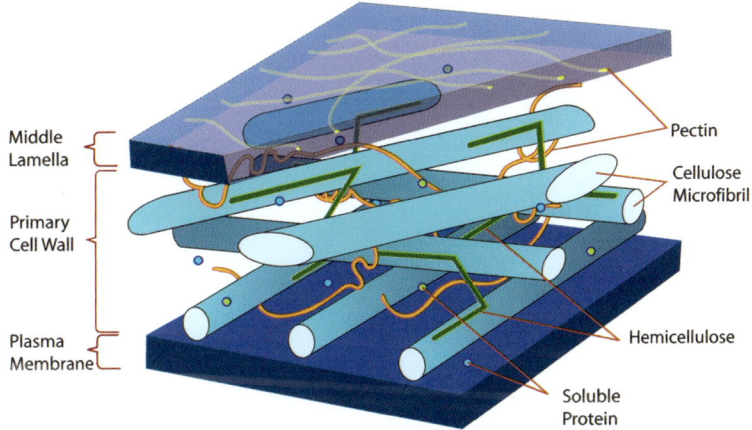

Typical plant cell wall showing the cross-linked structure
that provides tremendous strength to the plant.

Biodiesel produced from soybean oil.

As an alternative to the use of plant biomass, some biotechnology companies are developing large algae farms to produce biofuels such as vegetable oil, biodiesel, bioethanol, biogasoline, biomethanol, biobutanol, and potentially others as well. Algae basically use sunlight to produce lipids (fats) or oil. The advantages of using algae farms for energy is that they do not require fresh water (ocean and wastewater may be used) and they are generally harmless to the environment. Algae also grows rapidly – they can double their mass several times a day and can produce more oil than traditional oil crops such as soybean, rapeseed, and palm. Several companies and government agencies are investing in efforts to make this process more efficient and cost-effective at large scale. In addition, seaweed is being investigated as a renewable source of biomass for ethanol production.

Biofuel production in algae. The droplets are oils that can be used as fuel.

Biotechnology applications in the energy industry also include the generation of cleaner burning coal and petroleum products by removing the sulfur with sulfur-eating bacteria before their use as fuels.

Summary

Given the limited fossil fuels on Earth, alternate energy sources that utilize biological and renewable resources are becoming increasingly important. The use of bacteria, yeast, or plant sources to generate energy has the potential to eventually replace fossil fuel, coal, and natural gas, along with renewable wind, solar, and hydrodynamic technologies. Bioenergy and biofuels will continue to be a very important aspect of biotechnology today and in the future.

Concept Reinforcement

1. What is the difference between bioenergy and biofuel?

2. What are some common sources of biomass for biofuel production?

3. Besides ethanol, what other types of energy can biomass be converted into?

Section 3.11 – Genetically Modified Organisms

Section Objective

- Describe and discuss genetically modified organisms

What are Genetically Modified Organisms?

In a **genetically modified organism** (GMO), genetic information has been altered in some way. Usually this alteration is the addition of a new DNA element, or sequences, by utilizing modern recombinant DNA techniques, also called genetic engineering. However, the modification may be the deletion of a gene or set of genes in an organism. The new DNA sequences added are often from a different organism and confer upon the GMO an advantageous or beneficial trait. The organism involved can be a plant or an animal, but is most commonly a plant that is grown as a crop for food, oil, sugars, or fibers. Animals that are modified are often called transgenic organisms, particularly if new DNA from a different organism was inserted into the animal. In February of 2009, the United States Food and Drug Administration approved the first human biologic drug produced from a genetically engineered goat. The drug, called ATryn, is an anticoagulant that is used to treat blood clots during surgery that is purified from the transgenic goat's milk.

> **Genetically modified organism (GMO)**
>
> Organism whose genetic material has been altered through genetic engineering; often in reference to modified plants and the modification confers an advantageous trait to the organism.

Food crops are often modified, particularly corn and soy, because there is scientific evidence that the genetic modification will create a better plant, either for purposes of greater yield, better nutritional value or flavor, or better storage. GMO corn often contains a gene that allows farmers to use an herbicide that will kill the surrounding weeds but not harm the corn plant. This is the modification in Monsanto's RoundUp Ready corn, which is resistant to the herbicide RoundUp (the use of which allows for higher yield per acre). Currently in the corn market, Monsanto's triple-stack corn is the market leader. It contains three different genes: one to confer resistance to the herbicide RoundUp, and two others to confer resistance to the corn borer and rootworm insects. Rice has been modified to produce vitamin A, which increases the nutritional value of the rice and gives it a characteristic yellow appearance (and thus the name "Golden rice").

Chemical structure of glyphosate (the herbicide RoundUp).

GMO tomatoes have been engineered to have a gene that prevents them from freezing in a cold temperature or ripening until they reach the store. There are other examples of GMO crops as well, including soy plants and cotton plants that have been modified to be resistant to a specific insect pest that eats their leaves. This is accomplished by introducing into their DNA a gene that makes a protein that is toxic to the insect. A plum plant has been engineered to be resistant to a virus that commonly infects it (plum pox virus).

Genetically modified plum that is highly resistant to plum pox virus.

Over 56 different genetically modified crops have been tested in 34 different countries, highlighting the global significance of GMOs. Currently five countries produce >95% of the world's GMO crops, including the United States, Argentina, Brazil, Canada, and China. Other countries also produce GMO crops, but to a much lesser extent (Australia, South Africa, Mexico, Spain, France, Bulgaria, Romania, Uruguay, and Germany). Accounting for >50% of global acreage, soybeans continue to be the most widely used genetically modified crop.

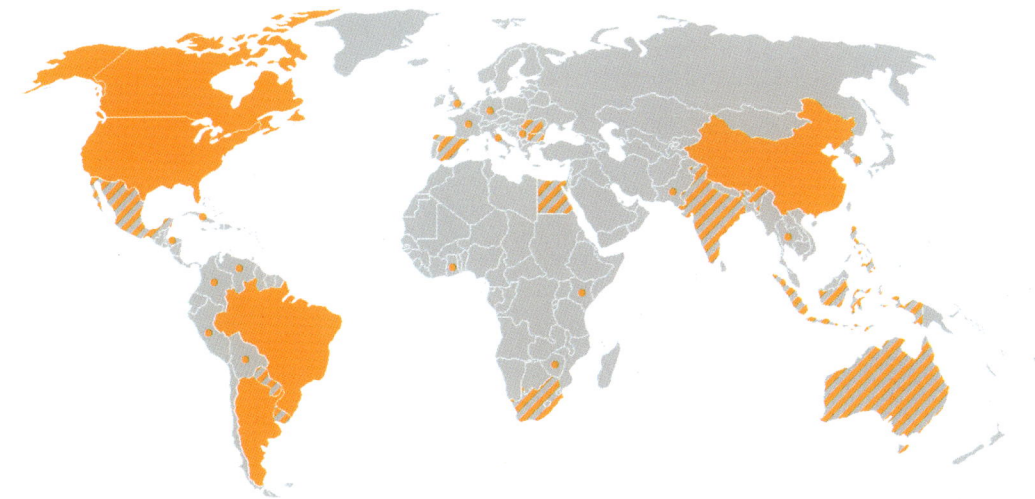

World map highlighting GMO production as of 2005. Solid orange indicates the top 5 countries producing more than 95% of the world's GMO crops. The striped orange indicates the other countries producing GMO plants.

Why and How are GMOs Created?

As discussed, many GMOs are created due to some benefit conferred by the genetic change and thus introduction of a new protein in the plant. Benefits include resistance to insect pests, herbicide resistance, increased yield, enhanced flavor or nutritional value, and other things as well. How are GMOs created? First, researchers might identify a trait that would be beneficial to have in a certain plant or crop. They then identify and isolate the gene, or genes, responsible for conferring this trait. Using biotechnology and recombinant DNA techniques, they amplify the gene and create a **transgene cassette**. This transgene cassette is made up of a promoter, the gene of interest, a terminator and a **marker gene** or resistance gene. The strong promoter drives expression of the target gene by binding of RNA polymerase to begin transcription and the terminator element stops transcription. The marker or resistance gene will express a protein that will make an organism that has incorporated the cassette resistant to a toxic compound. Thus, the purpose of the marker or resistance gene is to allow for the plants that contain the transgene to be selected for. In addition, the marker gene may be a protein that can be easily be detected. These elements allow the researcher to express the protein in the organism, verifying that it was incorporated, and potentially track its expression. One marker protein used in the generation of GMOs is a protein called GFP, or green fluorescent protein. This protein will give off light under the right circumstances. By having the target gene (whose activity usually can not be easily detected, especially if the protein is enhancing the flavor of the modified crop) linked to the marker and/or resistance gene in the transgene cassette, you can track the target gene by the expression and presence of the marker gene.

> ### Transgene cassette
>
> The genetic material inserted into a GMO; the transgene cassette minimally includes a promoter, transgene, terminator, and marker gene (or resistance gene).
>
> ### Marker gene
>
> The gene used to track transgenic material in a GMO; the marker gene often encodes for a protein that allows for resistance to a toxic chemical in the GMO that incorporated the cassette.

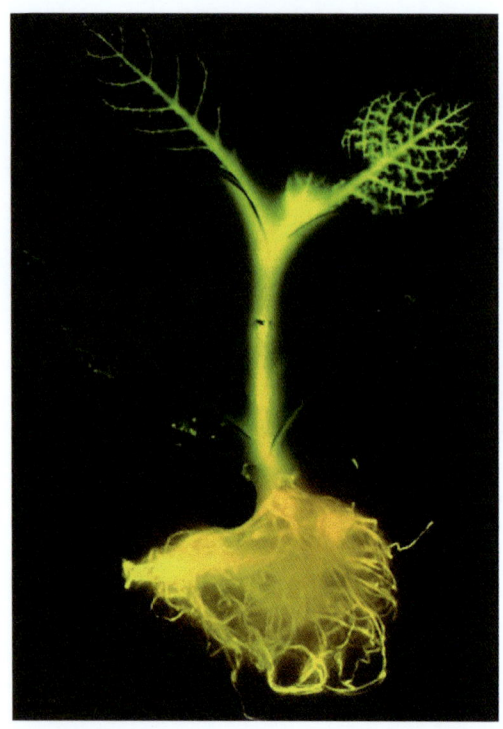

A transgenic tobacco plant that expresses green fluorescent protein (GFP) and thus glows green under the right circumstances.

Once the transgene cassette is made, researchers can put it into plants using a variety of methods. One is the use of a bacterium called *Agrobacterium* that can enter the plant cell and deposit its own genome into the plant cell DNA, along with the transgene cassette. Another method used to introduce foreign DNA into plant cells is the gene gun, which basically shoots small gold or tungsten particles coated with DNA into the cells. The DNA enters the plant cell nuclei and becomes integrated into the plant cell chromosomes. Exposing these plant cells to a chemical that would normally kill cells not containing the resistance gene allows for their selection. The presence of the resistance gene infers the presence of the target gene.

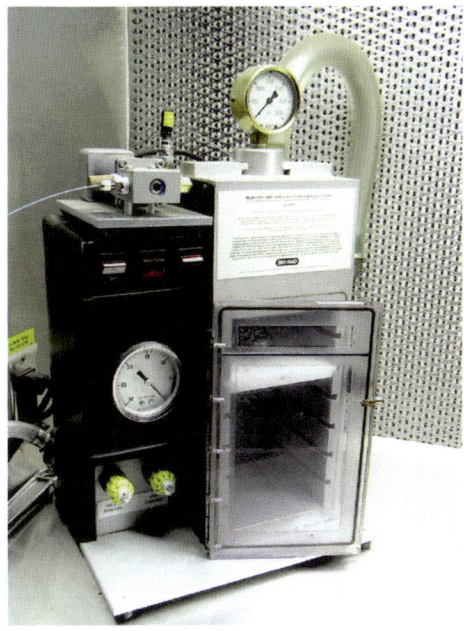

Typical gene gun used to create transgenic plants.

Controversial Aspects of GMOs

Even though there are many positive attributes of GMOs, there are also concerns related to their development and use. Will they harm the environment? Are there health concerns for GMO food crops? Countries have different regulations regarding the use of GMOs, with some significantly restricting GMO use. The European Union, for example, takes a very conservative stance when it comes to modifying foods. Simply stated, in Europe, it is judged that when it comes to food, practices must be proven safe before they are implemented. In the United States, generally speaking, the reverse is the case and practices must be proven to be unsafe or they can be utilized.

In addition, there are cultural differences with regards to how food is viewed. This plays an important role in the support or lack of support for GMOs. In Mexico for instance, corn is a sacred crop and manipulating it genetically is strongly frowned upon.

Some people believe that GMOs limit choice as they take the place of traditional crops. It is also difficult to predict what will happen to plants once we begin to create organisms with genes that cross the species barrier. Even though studies have not shown that GMOs are dangerous in conventional ways, some researchers are concerned about the effects of modifying plants on human and animal health and the environment. Will they alter ecosystems in any way? Many of these controversies are both scientific and ethical in nature, they are not easy to answer.

Summary

Genetically modified organisms have many different applications and the field of bio-technology has been very creative in their uses. They may be used in agricultural or medical research, in the production of biotherapeutics and other small molecule drugs, gene therapy tests, or simply as biological research tools to determine the fundamental processes that occur in both prokaryotic and/or eukaryotic cells and organisms. This is not without controversy though.

Concept Reinforcement

1. What is a genetically modified organism?

2. List the DNA elements in a transgene cassette and their function.

3. Name one genetic modification in a food crop and its advantage.

Section 3.12 – Genetic Identity and Short Tandem Repeats

Section Objective

- Explain how genetic identity is established using short tandem repeats

Genetic Identity and DNA Fingerprinting

DNA fingerprinting, which may also be referred to as DNA testing, DNA typing, or DNA profiling, is the technique used to distinguish individuals based on their genetic information. It highlights the relatively minor differences in DNA sequences between people, since these DNA differences are based on highly variable regions in the genome. These highly variable regions are within the noncoding regions of the DNA; that is, the DNA outside of the genes that code for proteins. This noncoding DNA makes up greater than 95% of the genome, with about 40-50% of it consisting of highly repeated regions whose exact purpose is unknown, but is under intense investigation.

There are a number of different techniques that have been used for DNA analysis for identity determination to distinguish one individual from another. These include **VNTR** and RFLP or PCR analysis, **STR** analysis, and Y-chromosome analysis.

VNTR Analysis

Variable number tandem repeats (VNTR) are regions in the genome in which a particular DNA sequence (usually 11-60 base pairs in length) is repeated a variable number of times between individuals. These regions can be analyzed by restriction enzyme fragment polymorphism (RFLP) analysis or by the polymerase chain reaction (PCR), which basically enable comparison of the number of repeats between different people because the resulting DNA fragments will be of different sizes. RFLP utilizes restriction enzymes that cut DNA at specific sites to make the VNTR segments available for Southern blot analysis, while PCR does the same thing much more easily and quickly. In Southern blot analysis, the DNA fragments are separated on an agarose gel by electrophoresis and then transferred to a supportive membrane. The membrane is then incubated (probed) with a radiolabeled piece of DNA that is specific for a particular region of DNA (one of the VNTR regions). Depending on the number of repeats a person has at that particular VNTR locus, the banding pattern will look different between individuals. By analyzing many VNTR loci, individual patterns can be determined that will be unique for each person.

DNA Finger-printing

Technique used to differentiate individuals based on difference in their DNA sequences; several different technologies are used for this purpose.

VNTR

Variable number tandem repeat; relatively short nucleotide sequences that are repeated at various locations in the genome; they often show variations in the number of repeats between individuals.

STR

Short tandem repeat; very short repeated DNA sequences (2-10 nucleotides in length, with 3-4 being the most common); they often show variability in the number of repeats present between individuals.

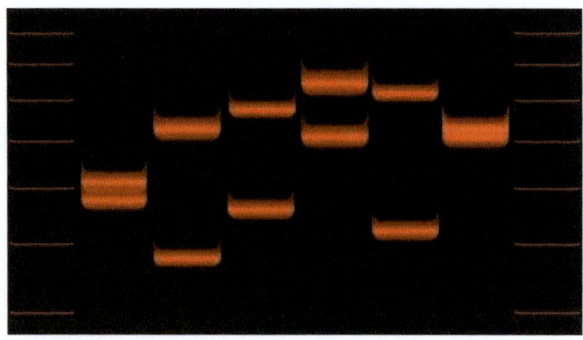

VNTR analysis showing different banding patterns between different individuals.

Y-Chromosome Analysis

Recent discoveries have identified highly variable regions on the Y-chromosome that allow for discrimination between male individuals. This technique uses short tandem repeats that are present on the Y-chromosome. Y-chromosomes are paternally inherited (i.e. from the father), so this type of analysis can help resolve cases of questionable paternity. **SNPs** may also be used for this type of analysis.

Short Tandem Repeats (STR Analysis)

The most common method used for human genetic identification today is STR (short tandem repeat) analysis. This technique became popular in the 1990s and has been used in many high profile criminal cases. STRs are very short repeated sequences (2-10 base pairs in length) found in the non-coding ("junk") DNA. Individuals will have varying numbers of repeats at each of the STR **loci** that are distributed across many different human chromosomes. The most commonly used loci have repeats of 4-5 bases and the number of repeats at any given loci can vary between ~5-35 depending on the genetic **locus**. These STR loci are analyzed by sequence-specific PCR (loci specific) and the resulting DNA fragments are labeled with different fluorescent tags during amplification. The different sized and fluorescently-labeled PCR products are then separated by **capillary electrophoresis** and detected using a laser. Although each loci may only vary once in every 10 to 20 people, by analyzing 13 loci or more, the power of discrimination to identify individuals is quite high (likelihood that two people will have the same exact STR pattern at 16 loci is 1 in ~10^{17}!). Thus this technique is incredibly powerful in identifying and differentiating individuals, since only ~$7x10^9$ people live on Earth today.

SNPs

Single nucleotide polymorphisms; a single base substitution of one nucleotide for another at a particular site in a particular gene.

Loci/locus

The specific position of a gene, DNA marker, or genetic marker on a chromosome.

Capillary electrophoresis

Movement of an electric field through a very tiny glass capillary containing gel matrix that allows the separation of DNA fragments that only differ in size by 1-2 base pairs.

Short Tandem Repeats (STR) are repetitive sequences:

Tetranucleotide
AAAG AAAG AAAG AAAG
Trinucleotide
CTT CTT CTT CTT CTT
Dinucleotide
AG AG AG AG AG AG

Example of short tandem repeats.

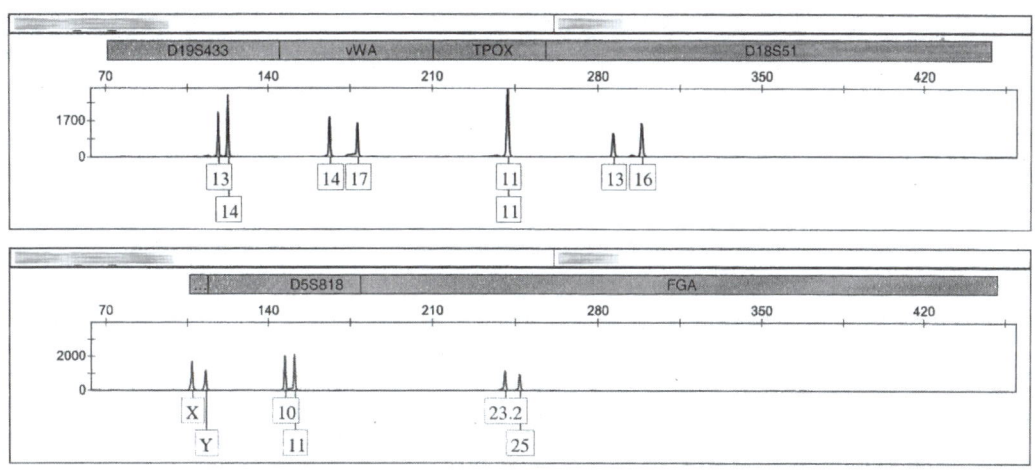

STR profile for an individuals DNA showing the results for seven different loci (D19S433, vWA, TPOX, D18S51, XY, D5S818, and FGA). Each peak on the electropherogram indicates an allele at that particular loci. This individual was female because of the present of peak at both the X and Y locus. The numbers indicated in the small boxes under each peak indicate the number of repeats at that locus.

Two companies make commercially available systems for STR analysis, Applied Biosystems Incorporated (ABI) and Promega Corporation. Each system analyzes the 13 core loci as determined by the Federal Bureau of Investigation (FBI), plus a few others for historical reasons. These loci are called the CODIS loci, which stands for Combined DNA Index System, which uses STR analysis and profiles to gather the DNA fingerprints from all convicted felons in the United States. The United Kingdom has a similar database called the National DNA Database (NDNAD), which is similar in size to the US database. In addition to genetic identity in forensics and crime scene investigations, STR analysis is also used for paternity cases. A child will have a profile that is a combination of the two parents. In addition, STR analysis is often used to identify victims in mass disasters or mass graves because it is such a sensitive and specific technique.

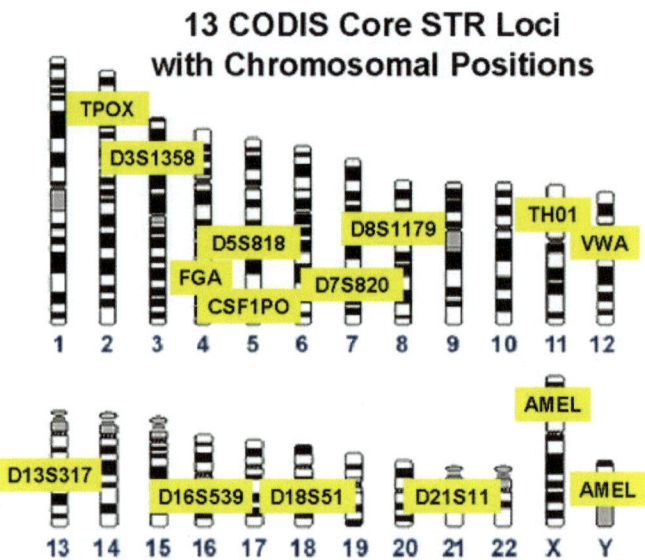

The 13 CODIS core STR loci with chromosomal positions indicated.

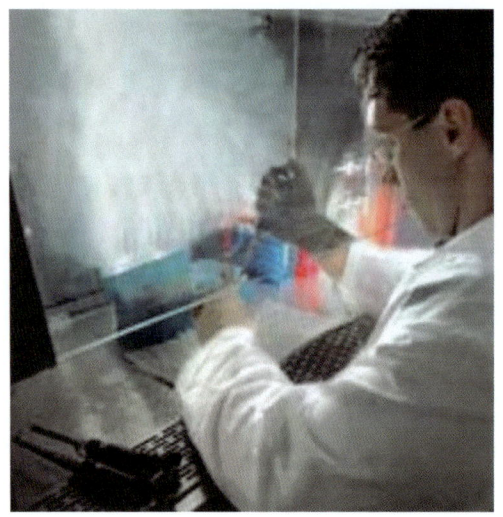

A scientist setting up a PCR reaction for amplifying the CODIS-required STR loci for DNA genotyping.

Ethical Issues in DNA Fingerprinting and Genetic Analysis

With the ease of obtaining and cataloging genetic information from numerous individuals, the question of confidentiality must be addressed. This is true for STR analysis. For example, although no diagnostic information is currently linked to STR analysis, it is possible that in the future that it might. Needless to say, there are many potential risks to privacy associated with the possible sharing of information with insurance companies, employers, other family member, etc. Any person's genotype is highly personal and its use in databases and research projects should be blinded so that a particular DNA fingerprint can not be traced back to any individual's name or contact information. As genetic information becomes more readily available, continued discussion of related ethical, legal, and social issues becomes increasingly important.

Summary

The analysis of genomic DNA has proved to be a powerful tool for human identification. The use of variable regions in the human genome, such as STRs, will undoubtedly find more uses in the future, as more and more information becomes available. Advances related to personalized medicine are a good example of this trend. Biotechnology has allowed for the development and commercialization of the technologies that allow for genetic fingerprinting, including VNTRs, STRs, or Y chromosome analysis. New and innovative technologies will expand these techniques in the future.

Concept Reinforcement

1. What is meant by the term DNA fingerprinting?

2. What is an STR and how are they used in genetic fingerprinting?

3. What is CODIS and what role to STR profiles play?

Section 3.13 – Mitochondrial DNA and Ancestry Tracking

Section Objectives

- Describe mitochondrial DNA and how it is used in ancestry tracking

Mitochondrial DNA Analysis

DNA fingerprinting takes advantage of repeated regions in the chromosomal DNA of human cells. For this type of analysis, the DNA needs to be in relatively good shape (that is, not destroyed or degraded). For highly degraded DNA samples, it is often possible to get genetic information from mitochondrial DNA. Each mitochondrion, of which a single cell possesses hundreds, contains a single circular piece of double-stranded DNA (mtDNA). Mitochondria are organelles within eukaryotic cells that are responsible for generating chemical energy (ATP) that the cell uses to accomplish many functions. Two hypervariable regions are amplified using **PCR** (HV1 and HV2) and then the DNA sequence is determined. Differences can be compared between individuals. Because mitochondria are only inherited maternally (i.e. from the mother), they can be used to compare relatives or relatedness. Two individuals who differ by one to two nucleotides in these regions are not considered to be related. Closely related individuals would have the exact same DNA sequence in these regions. The results of mtDNA testing are compared to same results of others to determine the time frame in which two people shared the most recent common ancestor.

The different mtDNA **haplotypes** appear to be continent-specific, with little mixing of mtDNA haplotypes from different continents. The oldest mtDNA haplotypes are found in Africa, with the newest being in the New World. This suggests that modern humans arose in Africa first, and then migrated around the world, arriving in the New World last.

> **PCR**
>
> Polymerase chain reaction; method for amplifying (copying) specific regions of DNA to many millions of copies.
>
> **Haplotype**
>
> A combination of alleles or SNPs at multiple sites (loci) that are inherited together as a group or block.

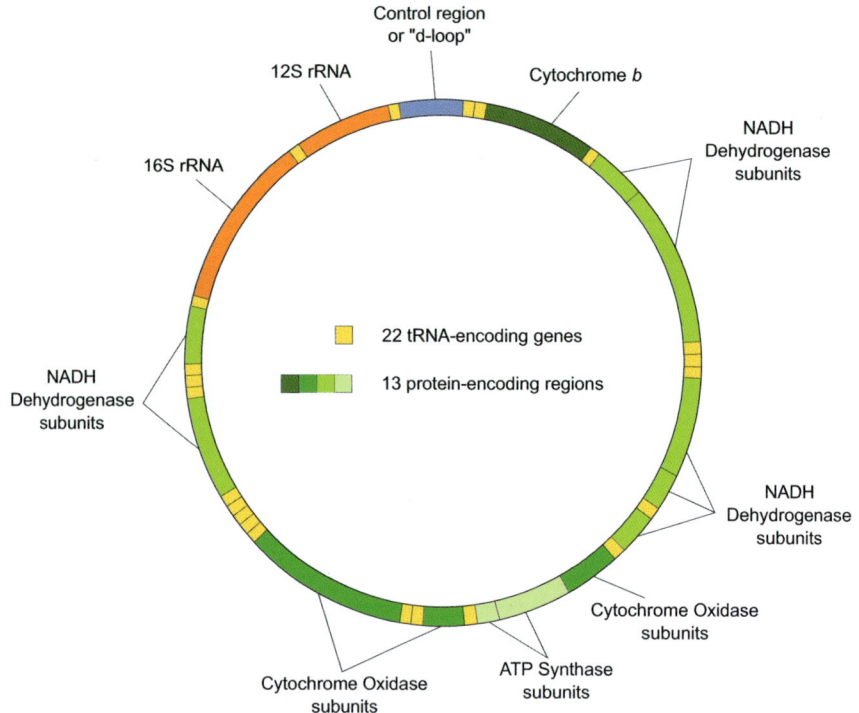

Human mitochondrial DNA is present as a single plasmid in each mitochondrion. It contains several different genes necessary for mitochondrial function, including protein translation and energy production through the electron transport chain.

SNP Analysis and the HapMap Project

Single nucleotide polymorphisms (SNPs) are single nucleotide differences at a particular site in a particular region of DNA. All versions may be observed in the general human population at varying frequencies. An example of a SNP would be an individual who has the sequence "GATCTGA" in a particular gene, while a different set of individuals has the sequence "GATCAGA." Thus, the polymorphism in this case is the "T/A" at the fifth base in the sequence. The SNP Consortium has a mission to determine and map about 300,000 evenly spaced SNPs within the human genome. It has been estimated that there may be as many as 10-30 million SNPs present in the human genome. Some are inherited in blocks, called haplotypes. Many haplotype SNPs have been transmitted through many generations without recombination, and can thus be used for ancestry and relatedness. Researchers are trying to associate different haplotypes with the prevalence of a particular disease in a population. Individual SNPs are being used in pharmacogenomics and personalized medicine, and this trend will only increase in the future.

The International HapMap Project is an organization whose primary goal is to develop a haplotype map of the human genome that will describe any common patterns of human genetic variation (i.e. SNPs). It is a collaboration between academic and private groups in Canada, China, Japan, Nigeria, the United Kingdom, and the US. The project had to sequence billions of base pairs of DNA of numerous individuals to identify more SNPs to be added to the database. The hope is that the information gathered will be invaluable to researchers investigating the genes affecting human health, disease, and the response to various drugs and environmental compounds. As modern humans spread throughout the world, the frequency of various haplotypes came to vary from region to region through random chance, natural selection, and other genetic mechanisms. Thus, any given haplotype can occur at different frequencies in different populations, particularly those that are widely separated and unlikely to interact. Mutations that arise create new haplotypes that may or may not spread throughout the world, depending on when they occur and how isolated the population is. All of the data generated by the project, including SNP frequencies, genotypes, and haplotypes, have been placed for public access on the International HapMap Project web site.

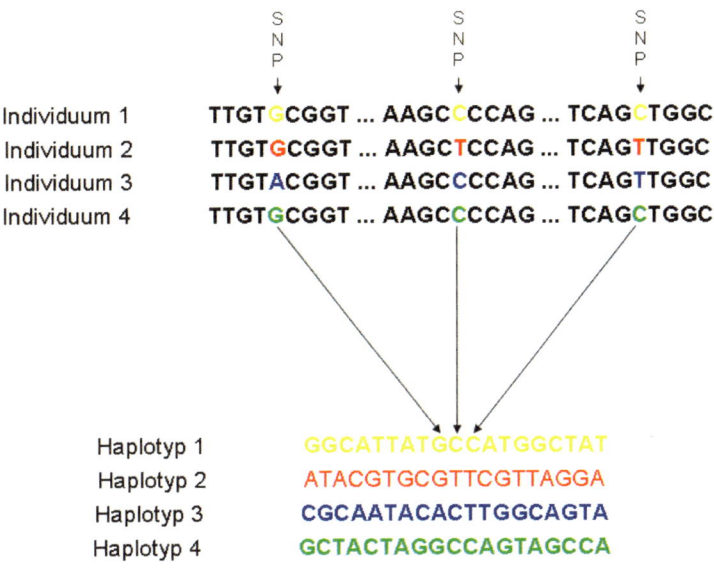

Different haplotypes from comparison of SNPs in a part of the same chromosome from four different individuals.

Ancestry Tracking

Genetic information can be used to track relationships between individuals and their ancestors, and thus is termed genetic genealogy. The two most common genealogy tests are Y-DNA analysis through SNPs or STRs present on the Y chromosome for the paternal line, or mitochondrial DNA (mtDNA) for the maternal line. These tests involve comparison of certain DNA sequences present on either the Y chromosome or in the mitochondrial DNA to estimate the probability that any two or more people share a common ancestor. This is done by estimating the number of generations separating the two individuals using a method developed by Bruce Walsh. The more similar the sequence, the more related the individuals or the closer in time. Y-STR analysis is used for recent ancestry, while Y-SNP analysis is used for ancient genetic ancestry.

Genealogical DNA testing methods are being used to trace the migration patterns of human populations throughout the world. For example, DNA testing has illustrated a remarkable link between the ancient Phoenician people and the present-day population on the island of Malta (off of the coast of Sicily). The Genographic Project is a five year research partnership launched by the National Geographic Society and IBM in 2005 to test the general public for either 12 STR loci on the Y chromosome or the HVR1 region in mitochondrial DNA. The primary goal of this project is anthropological, but it has also dramatically increased visibility of genetic genealogy. There are companies that focus on offering genetic genealogy tests to the general public (Genebase, OneGreatFamily, MyHeritage, KindredKonnections, TribalPages, 23andMe, etc.). Genebase gives the option for testing 44 Y-DNA markers in their service, while 23andMe offers personal DNA analysis, not just for ancestry tracking, but also information about health and disease traits. Such companies should clearly articulate to the individual consumer the results of the analysis, but also the limitations of the information that they are supplying. Genetic information always carries with it ethical concerns about health and disease traits, privacy, and familial relationships.

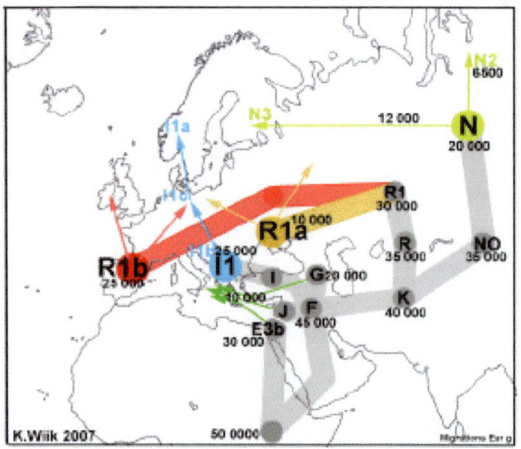

Map showing Y-chromosome haplotype migrations from Africa.

Summary

Variable DNA elements that can differ between individuals are being utilized for individual identification, but also for ancestry tracking to determine relatedness on a small scale for individuals, but also on a large scale to track population movements throughout human history.

Concept Reinforcement

1. What is mitochondrial DNA and how is it used in ancestry tracking?

2. What is the International HapMap Project?

3. Besides mitochondrial DNA analysis for ancestry tracking, what other DNA features are used for such a purpose?

Section 3.14 – Pharmacogenomics and Personalized Medicine

Section Objective

- Discuss pharmacogenomics and personalized medicine

Pharmacogenomics

It should not seem too surprising, since genes affect so many human characteristics, that genes and their protein products can affect the response of living organisms to environmental exposures, including drugs or nutritional supplements that we may take over the course of a lifetime. This effect of genotype on the response to drugs has been termed **pharmacogenomics** or **pharmacogenetics** and is the science that studies the inherited variations in genes that affect the response to chemicals. It can be thought of as the intersection between pharmacology/pharmaceuticals and genetics.

> **Pharmacogenomics or Pharmacogenetics**
>
> Interplay between genotype and response to various drugs and supplements; the science that studies these interactions.

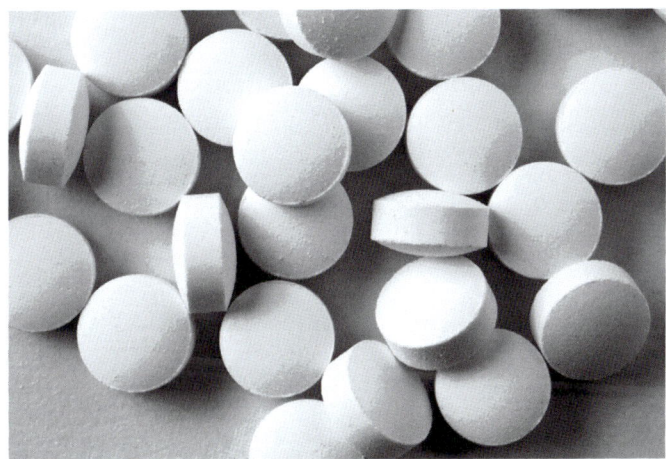

Pharmacogenomics focuses on predicting a patient's response to drug therapy based on their genetic makeup.

The goal of pharmacogenomics is to be able to predict whether a patient will have a good or bad response to a drug, or no response at all, to reduce the often-fatal adverse drug reactions (ADRs) that happen every day. It is believed that adverse drug reactions are a major cause of hospitalizations and deaths each year. Various proteins play roles in many areas of the response to a drug, including how the drug is absorbed, metabolized, excreted, and distributed throughout the body. These proteins and enzymes also affect how the drug interacts with the target, and thus its mechanism of action, efficacy, and toxicity. All of these variables make any person's response to a drug quite unique, with many differences among patients. It also makes any particular response unpredictable. To complicate matters even further, if a patient takes more than one drug at a time, very serious complications can occur because of the differences in how these proteins affect each drug independently and together. Such drug-drug interactions are difficult to predict and can can be quite harmful.

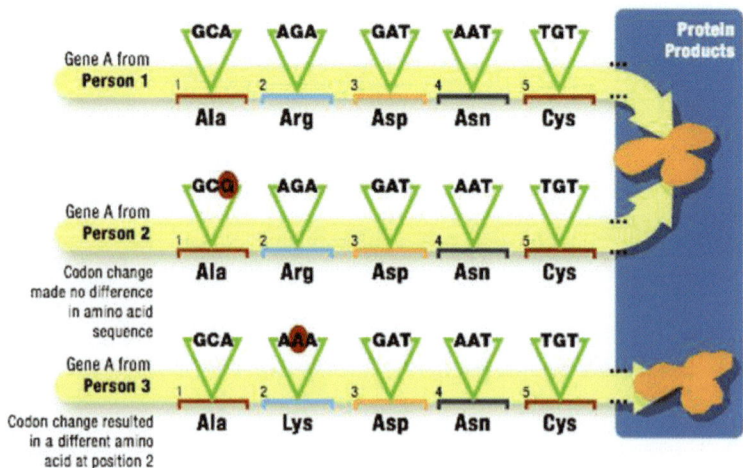

Differences in specific genes can affect the protein that is made
and how that protein functions in response to a drug.

Better ways to predict and detect such negative reactions to prescription and other drugs, as well as dietary supplements, is a key area in genetic and pharmacology research, and has been specifically funded by the National Institutes of Health (NIH) and other governmental agencies. The Pharmacogenomics Research Network, a consortium of scientists who are studying how genetic variation contributes to differences in drug responses, is one such group. The collected information about specific proteins, genes, and pathways is being integrated into the Pharmacogenetics and Pharmacogenomics Knowledge Base. It is also of key importance to pharmaceutical and biotechnology companies that investigate, design, and market new drugs. The current drug discovery process identifies drugs that affect the "average" patient, but in general no one drug is perfect for everyone, just like one shoe does not fit all. Ideally, a doctor would be able to screen patients for specific genes and based on that information determine which medication would be best suited for them and their particular medical condition. This is basically **personalized medicine**. Drug companies would prefer to screen patients prior to clinical trials so only those who might benefit would be included. This would also allow clinical trials to be smaller and faster, and thus much less expensive. In addition, diagnostic companies are very interested in creating new tests for those genetic elements determined to be important to drug responsiveness.

The response that a person has to any particular compound, whether it be a prescription drug, an over-the-counter (OTC) medication, or something found in the diet, is a complex trait that is influenced by many genes. So far it has been difficult to develop tests that will predict accurately such responses. There is now a race to identify and categorize differences in the DNA sequences of various genes that could potentially be used as a diagnostic tool to predict a patient's response to a drug. As discussed earlier, these nucleotide differences are called **Single Nucleotide Polymorphisms** (SNPs) and new techniques are emerging to identify and study them in many people. While human DNA is 99.9% similar between individuals, the 0.1% that is varies is primarily due to the presence of SNPs, which are single nucleotide differences at a particular site in a particular gene. These nucleotide differences in the DNA can in turn produce changes in the amino acid sequence of the protein, and potentially affect the protein's normal function,

Personalized medicine

The use of information from a patient's genotype or gene expression pattern to select a medication or therapy ideally suited for them; also can be used to predict which preventative measures might be most beneficial as well.

Single Nucleotide Polymorphisms

A single base substitution of one nucleotide for another at a particular site in a particular gene.

amount produced, or where it is localized in a cell or tissue. The differences in patients in these gene products that are involved in drug absorption (channels or receptors in the intestines), metabolism (**cytochrome P450 enzymes** in the liver), excretion (channels or receptors in the kidneys), distribution (carrier proteins in the blood), or the target protein itself for the drug can all affect the positive and negative response to any particular drug or compound. The most toxic drugs were found to be metabolized by the enzymes with the most genetic variability.

In addition to identifying which SNPs might affect drug responses, researchers are also trying to identify and catalog how gene expression levels may also be involved in this process. Having too much or too little of a particular mRNA, and thus protein, may dramatically affect the response, either positive or negative, to a particular drug or chemical.

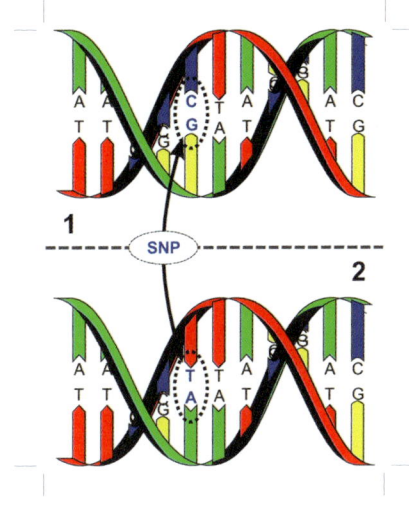

Diagram showing an example of a single nucleotide polymorphism (C/G versus T/A).

Personalized Medicine

The ultimate goal of modern medicine would be to accurately predict, and thus tailor, each patient's symptoms and disease to the proper drug regime, maximizing benefits and minimizing any toxic side effects. Current biomedical research has identified many genes that are involved in the response to various drugs and chemicals we might be exposed to. Many include enzymes involved in the break-down or inactivation of drugs.

In the past, determining the sequence of these SNPs for individual patients was difficult, time consuming, and expensive, and not really practical to do in a doctor's office, or even a hospital. New DNA analysis technologies, called DNA microarrays, may change this picture and eventually allow for SNP screening to occur in a doctor's office or hospital before a patient receives a medication. Such technologies could also be used to confirm the reason for an adverse drug reaction or drug-drug interactions. The advantage of DNA microarrays is that they allow for the simultaneous detection of many SNPs at one time (hundreds to thousands) in a relatively short period of time. The National Center for Biotechnology Information (NCBI) has developed databases for storing and annotating

Cytochrome P450 enzymes

A group of enzymes called cytochrome P450s oxidoreductases (CYP 450) are heavily involved in the metabolism of drugs in the body. They are a large family of related enzymes that are distributed throughout the liver, intestines, lungs, and skin, where they are located primarily in the mitochondria or endoplasmic reticulum of cells. They are also involved in metabolizing molecules normally found in the body. They generally act to increase the excretion of compounds by making them more water soluble. Many drugs or compounds may increase or decrease the activity or expression levels of various CYP enzymes and this is the major cause of adverse drug reactions or drug-drug interactions. One drug might inhibit the activity of a CYP enzyme required for the break-down of a second drug, thus producing toxic effects because of the prolonged exposure to the second drug. Naturally occurring substances found in foods can also alter the activity of various CYP enzymes, so drug-food interactions have to be closely monitored as well.

both SNP (dbSNP) and microarray data (GEO) that is accessible by researchers around the world.

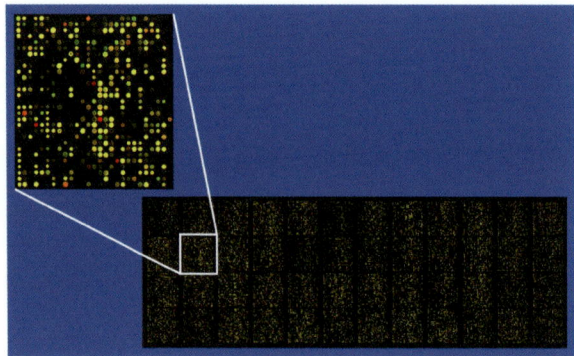

DNA microarray with thousands of individual spots for specific gene sequences.

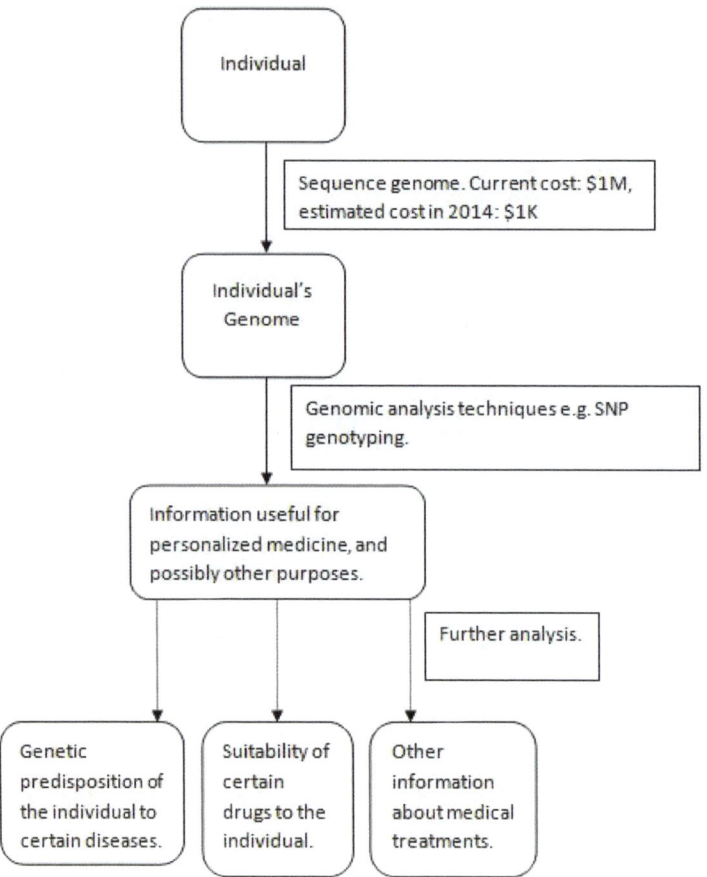

Overview of personalized medicine.

As discussed earlier, knowing a person's genetic makeup may also allow for early screening for diseases and the predisposition to various diseases, which will allow early lifestyle modifications and monitoring to minimize the effect of the disease. The promise of personalized medicine will make it possible to give "the appropriate drug, at the appropriate dose, to the appropriate patient, at the appropriate time." A high level of collaboration between scientists and physicians will be required to make the most of the potential of personalized medicine. To date, a few universities are developing educational opportunities in personalized medicine or something similar, including Duke University, Harvard, Mount Sinai Hospital in New York City, and Arizona State University. The Laboratory of Personalized Molecular Medicine was founded in 2007 to identify specific mutations in genes linked to clinical outcome in patients with leukemia and lymphoma (both cancers found in the blood).

Ethical Implications of Personalized Medicine

Although the idea of personalized medicine and the corresponding ability of physicians and drug companies to tailor drug delivery for each patient based on unique genotype sounds attractive, there are associated reasons for concern. For example: Who will have access to such information? Who will pay for the cost of determining each person's genotype? (currently only 5% of private insurance companies reimburse the cost of genetic tests.)

There are fears that those with specific genotypes will be discriminated against for medical care, health insurance, and/or employment. The federal government is working on legislation to address these concerns through legislation entitled the "Genetic Information Nondiscrimination Act."

In addition, a person's environment, diet, age, lifestyle, and health status will still influence a response to any medication and should always be taken into consideration. Genetic makeup is very important, but it is not the entire picture, and physicians, consumers, and researchers should always keep this in mind. The "Genomics and Personalized Medicine Act" has been introduced to overcome the scientific barriers, negative market pressure, and regulatory obstacles that have in the past stood in the way of better treatments.

Summary

Biotechnology has roles to play in pharmacogenomics and personalized medicine. Biotechnology companies are developing biotherapeutics, which could benefit from genetic information about individual responses to certain medications. They are also developing better diagnostic tests for the different genes and proteins involved in the differences in response to drugs or other compounds. These fields will change greatly over the next several decades and it will be interesting to see what emerges during our lifetimes.

Concept Reinforcement

1. What is pharmacogenomics/pharmacogenetics?

2. What are SNPs and how do they relate to pharmacogenomics?

3. What are cytochrome P450 enzymes and how to they relate to personalized medicine?

Section 3.15 – Drug Discovery and Vaccine Development

Section Objective

- Explain drug discovery and vaccine development

Modern Drug Discovery

The process of discovering and developing a new chemical compound or new biotherapeutic for use in treating human or animal diseases is a long and expensive process. The current estimate for the time it takes to develop a new drug is approximately 10-15 years, costing approximately 1 billion dollars! Most would admit though that it is well worth the investment, given the advances in medicine that result in health benefits for humans and animals. Many older drugs were thought of as the "low-hanging fruit" because they were easier to identify and develop, possibly because they were based on traditional plant remedies that had been passed down through generations, while newer drugs are targeting more complicated diseases and/or may be more difficult to manufacture.

The modern drug discovery process has many different steps, which are outlined in the two figures below. The first figure highlights the time it approximately takes for the major segments and also emphasizes the attrition (loss) of potential drug compounds as the process continues, starting with tens of thousands at the beginning, down to one at the end.

Pharmaceutical and drug companies possess very large collections (called libraries) of vastly different chemicals that they use to screen for activity against a particular target. That target has been researched heavily and identified as the cause or potential cause of the disease or medical condition that the drug is being developed for. This aspect of drug discovery, in which the specific molecular reason for a disease is determined, is often the most difficult step of the process and requires input from many different researchers and scientific disciplines. Once the target is identified, which is most often a specific protein in the body, companies will screen their compound libraries to see if any of the chemicals (or biological molecules in the case of biotechnology companies), can affect the activity of the target in the desired way. This may be to increase the activity of the target or decrease the activity of the target, depending on the disease and the role of the target molecule. The key point at this step is that the company or laboratory needs to have a way to measure the activity of the target, often requiring that the target protein is in a relatively pure form and there is a quantitative method for detecting it. This can be a very difficult process.

Once compounds are identified that affect the target (primary screening), they are screened again (secondary screening) to verify that they do indeed affect the target in the desired way and to see if they affect other proteins nonspecifically in ways that might cause toxic side effects. Medicinal chemists may take these promising compounds and make modifications to their chemical structures to either enhance their activity against the target (and thus hopefully their efficacy in the body) or decrease their side effects (and thus hopefully their toxicity). These enhanced compounds are then tested in animal studies for **a**bsorption, **d**istribution around the body, **m**etabolism, **e**xcretion, and **tox**icity (**AMDETox**). The studies are called preclinical studies and precede the use of the potential drug in humans (clinical trials). In human clinical trials, the potential drug is tested for all of the same things it was tested for in animals (ADMETox), as well as its effectiveness in treating the disease being investigated.

The United States Food and Drug Administration (FDA) oversees the entire process, as do similar agencies exist in other countries. They verify that all of the necessary studies were done correctly and that the potential drug has benefits that outweigh any side effects and risks. Before entering human clinical trials, the FDA has to approve an "Investigational New Drug" (IND) application for all New Chemical Entities (NCEs) or biologics, and before a drug is being sold on the market, must approve a "New Drug Application" (NDA) or a "Biologics License Application" (BLA) if the drug is a biologic drug (biotherapy or biotherapeutic). In addition, companies that launch new drugs are required to monitor what happens after patients start using it during what is called "Post Launch Marketing Surveillance." This allows for any side effects of the new drug to be tracked to try to avert any major adverse drug events, however, this surveillance is currently not mandatory for all drugs. Serious negative side effects do occur that were not caught in the original clinical trials, partly because millions of people can not be used as participants in clinical trials and rare events only occur following release and wide spread use. For example, Vioxx was withdrawn from the market due to adverse effects on the heart that were only detected following administration to numerous patients.

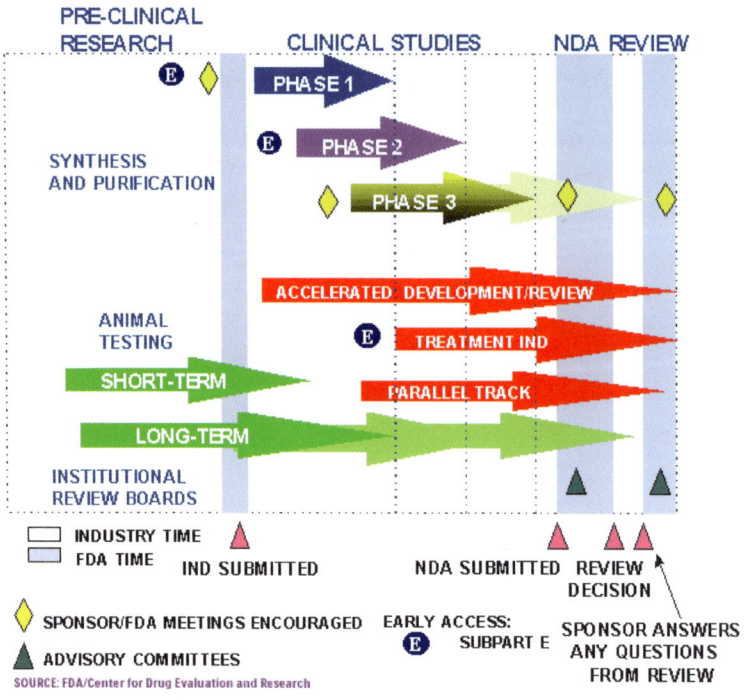

Overview of the drug discovery process highlighting the number of compounds that do not make it through the screening process and the amount of time it takes.

The Drug Discovery, Development and Approval Process

It takes 12-15 years on average for an experimental drug to travel from the lab to U.S. patients. Only five in 5,000 compounds that enter preclinical testing make it to human testing. One of these five tested in people is approved.

	Discovery/ Preclinical Testing		Phase I	Phase II	Phase III		FDA		Phase IV
Years	6.5		1.5	2	3.5		1.5	15 Total	
Test Population	Laboratory and animal studies	File IND at FDA	20 to 100 healthy volunteers	100 to 500 patient volunteers	1000 to 5000 patient volunteers	File NDA at FDA	Review and approval process		Additional post marketing testing required by FDA
Purpose	Assess safety, biological activity and formulations		Determine safety and dosage	Evaluate effectivenes look for side effects	Confirm effectiveness, monitor adverse reactions from long-term use				
Success Rate	5,000 compounds evaluated		5 enter trials				1 approved		

Source: Pharmaceutical Research and Manufacturers of America, www.phrma.org

The drug discovery, development, and approval process in the United States. FDA = Food and Drug Administration; IND = Investigational New Drug; NDA = New Drug Application. A BLA (Biologics License Application) is comparable to an NDA but for a biotherapeutic.

The development of a biologic drug is very similar to a standard small molecule (chemical) drug, but the disease targets, manufacturing process, safety tests, and mechanisms of action in the body may be different. Biotherapeutics (therapies or drugs based on biological compounds) are almost always proteins. They include growth factors like human growth hormone to treat small stature or other growth abnormalities, hormones like insulin to treat diabetes, erythropoetin (protein that enhances red blood cell formation) for anemia, enzymes for cystic fibrosis or severe combined immunodeficiency (SCID), and monoclonal antibodies to treat a variety of conditions including asthma, cancer, arthritis, and Crohn's disease.

As discussed earlier, the production of human **monoclonal antibodies** has become an incredibly important contribution that biotechnology has made regarding medicine and disease treatment. Monoclonal antibodies are made by B-lymphocytes of the immune system and recognize a specific aspect of a foreign molecule or organism (antigen). There are currently at least 25 different antibodies on the market to treat such diseases as asthma, rheumatoid arthritis, and cancer. This number will continue to increase in the future as biotechnology and drug companies devote more research and development to identifying novel therapeutic antibodies or proteins.

Vaccine Development

Recombinant proteins and nucleic acids are also used in **vaccines** to elicit an immune response to a particular pathogen, such as a virus or bacteria. Edward Jenner was the first to discover that when people were administered cowpox they became immune to human small pox. This was the first example of a vaccination and many more vaccines have been introduced in the 1800s and 1900s for a wide variety of pathogenic organisms or toxins.

> **Monoclonal antibody**
>
> Antibody that recognizes a single epitope on an antigen; derived from a single B cell and its daughter cells.
>
> **Vaccine**
>
> A biological preparation or biologic molecule that elicits an immune response to a particular disease or infectious agent or improves immunity to a particular target.

259

One of the most important vaccines developed was that of the polio virus. In the early 1950s, there were approximately 100,000 cases of polio in the US. This vaccine, developed in 1955 by Jonas Salk, consisted of an inactivated (killed) poliovirus. This vaccine was injected, but a later vaccine developed by Albert Sabin using attenuated (live but greatly weakened) virus could be given orally. Most vaccines currently in use are injected, though the oral polio vaccine is still in use today.

Jonas Salk, inventor of the first polio vaccine.

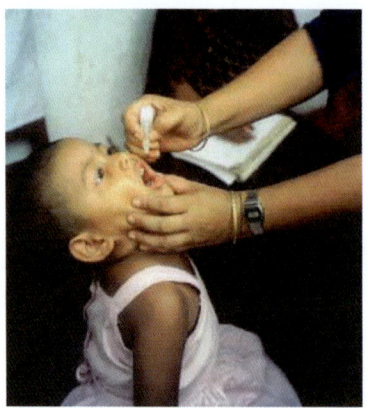

Child receiving the oral polio vaccine for immunization against polio.

In the past, inactivated whole organisms were used in the development of vaccines (such as polio), but now single pure viral or bacterial proteins, or mixtures of pure viral or bacterial proteins, can be used to develop vaccines. This allows for safer vaccines, since no whole organisms are involved, even if the organisms are live, attenuated (weakened), or dead. This takes advantage of the tools of biotechnology to manipulate DNA, and thus proteins, to engineer the target molecules of choice for inclusion in the vaccine. Vaccines for Hepatitis B and Human Papilloma virus (HPV) are both recombinant protein subunit vaccines.

DNA has the potential to be used in a vaccine, as the protein(s) antigens can be cloned into a plasmid using recombinant DNA techniques. Then, the plasmid DNA itself can be introduced into an animal or human. A single recombinant DNA vaccine has been approved for use in horses against West Nile Virus and an avian (bird) flu vaccine for humans is in development.

Reverse genetic approaches are being used to design a recombinant flu vaccine in which the antigenic (immune response producing) genes from a pathogenic strain of flu (strain 1) are combined with the necessary genes to make a complete virus from a nonpathogenic strain of influenza (strain 2). This recombinant virus is grown in animal cells and the recombinant viral particles purified and used in the vaccine. The influenza viral particles will not cause any disease, but since they contain antigens from the pathogenic strain the body can illicit an immune response to provide protection against the pathogenic strain (strain 2).

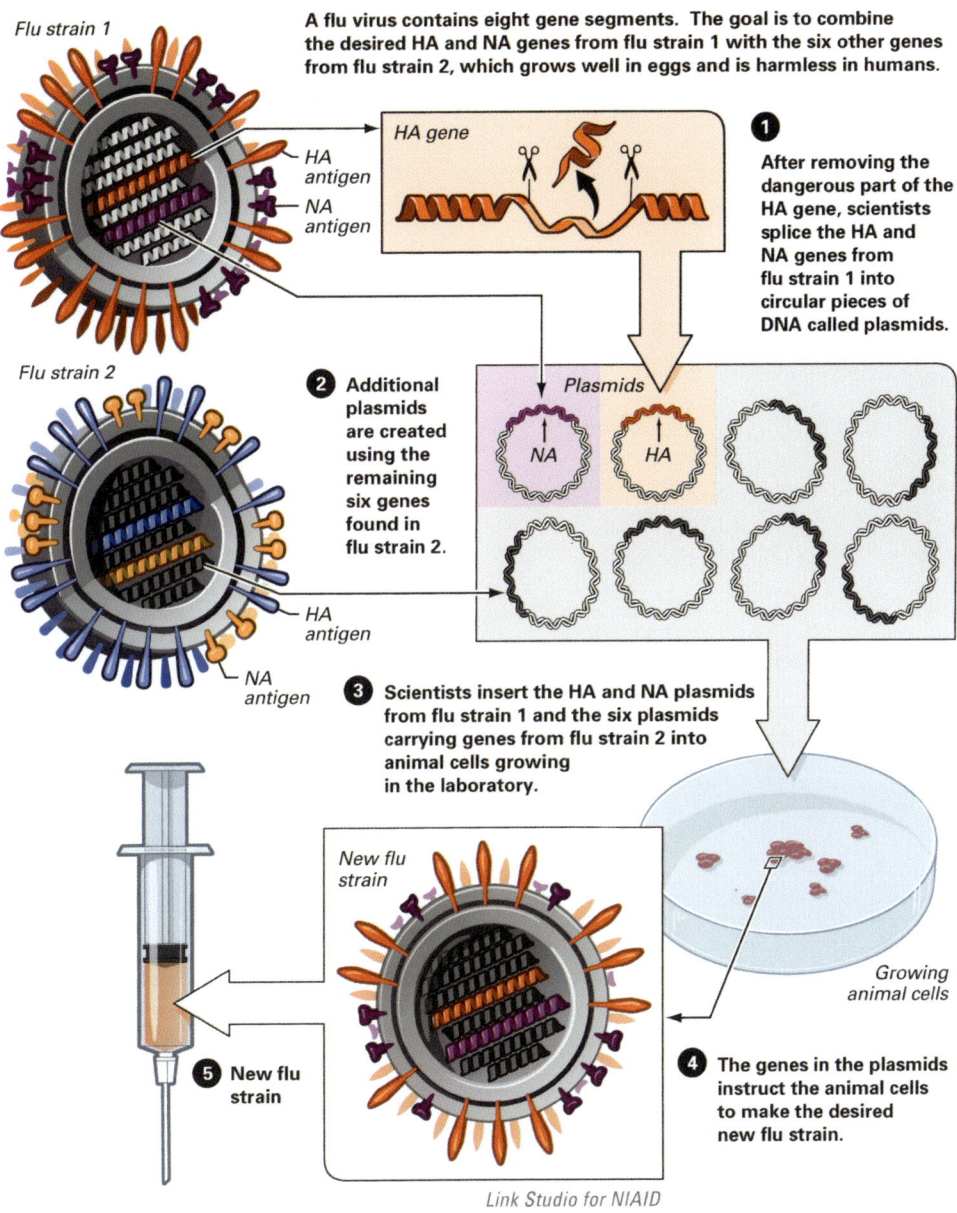

Flu strain 1

A flu virus contains eight gene segments. The goal is to combine the desired HA and NA genes from flu strain 1 with the six other genes from flu strain 2, which grows well in eggs and is harmless in humans.

HA antigen
NA antigen

HA gene

1 After removing the dangerous part of the HA gene, scientists splice the HA and NA genes from flu strain 1 into circular pieces of DNA called plasmids.

Flu strain 2

2 Additional plasmids are created using the remaining six genes found in flu strain 2.

Plasmids

NA   HA

HA antigen

NA antigen

3 Scientists insert the HA and NA plasmids from flu strain 1 and the six plasmids carrying genes from flu strain 2 into animal cells growing in the laboratory.

New flu strain

Growing animal cells

5 New flu strain

4 The genes in the plasmids instruct the animal cells to make the desired new flu strain.

Link Studio for NIAID

Vaccine development plan to create an immunization against avian flu, using a reverse genetics (biotechnology) approach.

Summary

Biotechnology is playing an ever-increasing role in drug and vaccine development. It is providing many avenues of research and development for new biotechnology companies, in addition to established pharmaceutical companies. Increasing efforts are being put into developing drugs for difficult diseases such as malaria and for developing a vaccine to HIV to combat AIDS.

Concept Reinforcement

1. Outline the modern drug discovery process.

2. What role does the FDA play in drug development?

3. What role does biotechnology play in vaccine development?

Appendix

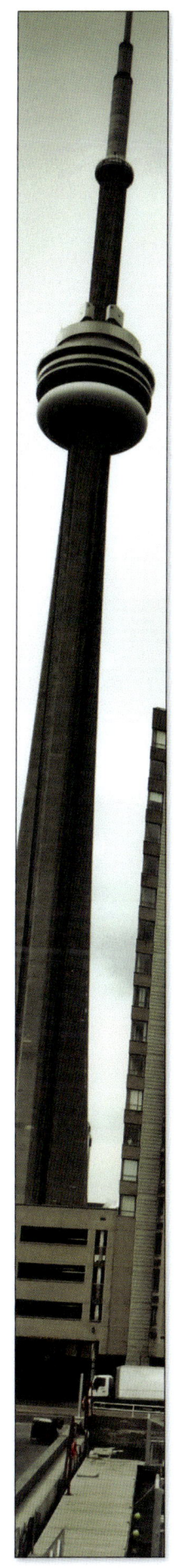

A Beginner's Guide to Biotechnology Unit 1

Section 1.1

1. Biotechnology has had a remarkable impact on the world we live in. From the foods we eat (cheese, corn chips, beer, soy products), which almost all contain genetically modified components, to pharmaceuticals that are manufactured using recombinant DNA technologies, biotech has had an important effect, and the impact will continue to increase in the future.

2. Phenotype is another word for physical trait, or how something looks. Phenotypes might include an organism's color, pattern, shape, size, number of limbs, or other physical characteristics.

3. DNA (deoxyribonucleic acid).

Section 1.2

1. Traditional biotechnology utilized whole organisms for human use, while modern biotechnology utilizes not only entire organisms, but their parts and processes at the molecular level. In addition, modern biotechnology allows for genetic manipulation of organisms for many different applications, including agricultural, medical, and industrial.

2. Molecular biology is the isolation and study of the individual biological molecules in a cell or organism, their functions, and how they interact together to accomplish complex physiological processes.

3. Biotechnology takes advantage of the information and advances in molecular biology to find useful applications for them in agriculture, medicine, and industry.

Section 1.3

1. The products of biotechnology are very complex and often incorporate elements of other areas to be successful. Math, chemistry, physics, engineering, and computer science are critical for most biotechnology applications.

2. Genetic modification means that the DNA of the organism has been altered. Usually, it is altered in a way that confers some desirable trait, such as resistance to disease or blight, or the ability to resist frost or rot. The gene for that desired trait it often obtained from an organism different than the plant that it is placed into.

3. Examples include enhance nutritional value by adding or increasing production of essential amino acids, vitamins, or beneficial lipids (such as omega3 fatty acids) in foods, enhanced drought or salt tolerance to allow for crops to grow in harsh environments, modifying plants to remove toxic wastes, or modifying plants to produce byproducts that could be used as alternative energy sources.

4. The field of biotechnology arose through the efforts and collaborations of many scientists distributed throughout the world. The fruits of this research have impacted people in many countries through modified crops and biotechnology drugs (biotherapeutics), to name just a few.

Section 1.4

1. The issuance of a patent allows a company or person to exclude others from using or taking advantage of their invention.

2. The US Food and Drug Administration (FDA).

3. The Recombinant DNA Advisory Committee (RAC), which is part of the NIH.

4. GCP stands for Good Clinical Practice, and ensures clinical trials with human subjects will be conducted safely, ethically, and effectively.

5. Critical components of an SOP are: reagents (chemicals, etc.), and equipment required, protocol (recipe) for making the product, who is making the product and when, confirmation that all of the reagents used are within their expiration dates, how the product will be tested to verify it is of high quality, and how the product will be stored and for how long (expiration date).

Section 1.5

1. There are many challenges facing modern global biotechnology, some include public and governmental education, access to biotechnology and development of biotechnology in developing countries, and the scientific hurdles facing biotechnology for the development of new disease treatments or for new biofuels.

2. While there are many opportunities facing modern global biotechnology, including the promise of bioprospecting and genetically modified organisms, and the ability of biotechnology to aid local regions in developing countries with solutions tailored to the particular needs of the local community, which compliment their existing human and environmental resources.

3. Bioprospecting is the search for novel biomolecules from around the world for new uses, such as medicines or industrial compounds. These novel biomolecules are often found in extreme and remote environments, particularly ocean environments.

Section 1.6

1. A good experimental design should contain a clear objective for the experiment, an understanding of which variables will be measured and which will be held constant, how many replicates should be performed and what are the proper controls, how the data/results will be analyzed, and what errors might occur.

2. Innovation and creativity, with an eye for practical applications of basic research.

3. The purification and analysis of nucleic acids and proteins.

Section 1.7

1. The two major differences between the research laboratory and the production laboratory are the amount of regulations and the scale. The production laboratory is much more highly regulated, requiring more documentation (paperwork) and monitoring, and usually performs techniques at a much larger scale.

2. A CMO, or contract manufacturing organization, is a company whose expertise is in manufacturing products for their customers. Often these products are biological molecules, since the manufacturing processes for these types of molecules is so complex and requires lots of special expertise and very expensive equipment.

3. Good manufacturing practice (GMP).

Section 1.8

1. Probiotics are live cultures of intact microorganisms (usually bacteria and/or yeast) that are included in food or food supplements to enhance the health of the host taking them. Biotherapeutics are drugs or medical treatments that are complex biologic molecules that are purified directly from a biologic source or expressed and purified from a host biologic source. Nutraceuticals are a general category of dietary supplement or food additives implied to have a pharmaceutical benefit

2. Biosimilars are the generic version of biotherapeutics that may not be identical chemically to the original drug, but must show comparable therapeutic actions to be deemed safe and efficacious for human use.

3. Monoclonal antibodies.

Section 1.9

1. Latex does not protect from many different chemicals because of its pore size. In addition, some people can become allergic to latex.

2. Safety goggles or glasses, non-latex gloves, and a laboratory coat.

3. Make sure the area is well ventilated. The dry ice should be contained in an ice bucket or foam container so that it does not freeze and crack the surface that you are working on. You should not seal dry ice – because it sublimates, it will give off gas and could cause an explosion if placed in a sealed container. In addition, you should protect your hands or other exposed skin from the extreme cold of dry ice with insulated gloves and clothing.

Section 1.10

1. There are four different biosafety level designations. They range in severity from one being not hazardous or minimally hazardous to level four being the most extreme hazard level.

2. An MSDS, or Material Safety Data Sheet, is information specific to each product or ingredient in a product that contains all safety information concerning that chemical or compound.

3. First, you should call a safety officer to the lab. Even though you might have a good idea of what the liquid is, you can't know for sure because one of the carboys is unlabeled, and you can't be sure which one spilled. If the substance can be identified, the MSDS should be consulted in order to determine how to proceed with the cleanup. Always follow established guidelines for handling hazardous chemicals and always use personal protective gear.

Section 1.11

1. The general public must have trust and faith in the scientific community or the information gained will not be applied to practical problems that can benefit all.

2. Plagiarism is using words, phrases, or the ideas of others as if they were your own. It is a form of scientific misconduct, along with falsifying or misrepresenting results, unwillingness to share information and scientific tools, and lying about results or methods.

3. Besides plagiarism, falsifying data for presentation or publication, failing to share results and reagents when possible, or lying about results are all considered scientific misconduct.

Section 1.12

1. Experimental design is the complete plan for conducing an experiment, including what question to ask, what reagents and equipment will be needed, what samples types and how many replicates to test, what controls will need to be performed, and how the results/data will be analyzed.

2. Precision is how close replicate measurements from the same sample are to each other, while accuracy is how close a given measurement is to a established standard.

3. A product team in biotechnology may involve all of the following members: research scientist, manufacturing scientist or engineer, marketing specialist, sales force, packaging and shipping expert, legal representative, QC or QA scientist, and/or a bioinformatics expert.

Section 1.13

1. Effective communication involves knowing the purpose of the communication and the audience. It involves knowing the types of presentation tools available, as well as the proper use of text and graphics. Tell the audience what they are going to learn, tell them, and then review with them what you covered. Make sure the communication tells a story and leads the audience from beginning to end, regardless of the purpose (entertaining, informing, or persuading).

2. Story-boards are a tool that can be very helpful when designing, planning, and generating an oral presentation. They assist in laying out and defining the ideas that will be covered, the figures that might be used for each point, and the transitions between slides. They provide a framework in which the detailed presentation can be built.

3. PowerPoint can be a very effective tool if used properly. Its advantages include the ability to include graphs, movies, sound, etc., along with textual information. It can be easy to view for an audience and paper copies can be provided as a resource for listeners. Disadvantages include that it can be distracting if too much text or other information is included per slide, if the speaker is not familiar with the topic or presentation and simply reads the slides, or if the purpose of the presentation does not match the advantages of PowerPoint.

Section 1.14

1. R&D scientists work at the laboratory bench developing new ideas for potential technologies or products. Manufacturing or production scientists actually produce those products on a large scale for sale to customers. While both positions can be obtained with a bachelor's degree in a scientific field, a master's degree or PhD degree are preferred and more common.

2. Quality control, quality assurance, and regulatory affairs scientists are all involved in most biotechnology companies and play key roles in verifying and maintaining product quality.

3. Yes, a number of universities and colleges have Master's and PhD degrees available in biotechnology.

Section 1.15

1. Sales, marketing, law, finance and accounting, customer service, information systems support, and packaging and shipping.

2. A patent agent is a person who has passed a registration exam and is able to prepare, file, and prosecute patent applications with the United States Patent and Trademark Office. They are not a lawyer and do not have a degree in law.

3. Some biotechnology companies have set up public outreach programs that help educate children and adults about the science of biotechnology and its uses.

A Beginner's Guide to Biotechnology Unit 2

Section 2.1

1. a. 160 amu
   b. 255 amu
   c. 342 amu

2. 6.022×10^{23} molecules.

3. a. 1 molar; 1 normal
   b. 1 molar; 1 normal
   c. 1 molar; 2 normal
   d. 1 molar; 1 normal

Section 2.2

1. The metric system is a system of measurement based on the use of decimals; different base units are used, but all based on multiples of 10 with different prefixes to differentiate them. The SI system was based on the metric system but only uses seven base units. See the tables in Section 2 for details.

2. The metric system is a system of measurement based on the use of decimals; different base units are used, but all based on multiples of 10 with different prefixes to differentiate them. The SI system was based on the metric system but only uses seven base units. See the tables in Section 2 for details.

3. a. 12.037g
   b. 4.00g
   c. 7.454g
   (For a 0.1M solution you would need 0.1 molecular weight of the compound in grams. The molecular weights of the compounds listed are 120.37, 40.00, and 74.54g)

4. For 1 liter of a 1M solution of KCl you would add 74.54g to 1L of water. However, if you are only making 100milliliters (1/10 of a liter) of the 1M solution you would only need 1/10th of the amount of KCl, and thus 7.454 grams.

Section 2.3

1. Smallest amount of a substance that can be simply detected by a specific measurement technique (not necessarily reliably).

2. Smallest amount of substance that can be accurately quantitated by a specific measurement technique.

3. Smallest amount of a substance that can be reliably detected by a measurement process.

4. Equivalent to five times the minimal detectable limit (MDA) for any measurement technique; between limit of detection and limit of quantitation.

Section 2.4

1. Eukaryotic organisms contain a nucleus, while prokaryotic and archae organisms do not.

2. Actin and tubulin.

3. Energy production through aerobic respiration to produce chemical energy in the form of ATP.

Section 2.5

1. Archaea.

2. Archaea organisms can often survive very extreme and harsh environments, which may be useful in biotechnology for the isolation of enzymes or proteins that can withstand these harsh conditions.

3. Escherichia coli (E. coli).

4. Gram+ bacteria will stain with crystal violet dye while gram- will not. This is due to differences in the cell wall structure of the two types of bacteria.

Section 2.6

1. Eukaryotic cells contain a nucleus and other membrane-bound organelles, tend to be much more complex, and have lower metabolic rates and doubling times as compared to prokaryotes. Prokaryotic organisms are usually single-celled while eukaryotic organisms are multi-cellular. Eukaryotic RNA synthesis and protein production occur in separate compartments in the cell, while in prokaryotic cells they do not.

2. Plant cells, although eukaryotic, have outer cell walls that animal cells do not and also contain plastids/chloroplasts for photosynthesis.

3. Cellulose, hemicelluloses, and pectin. Chitin. These molecules provide structural support to the plant and fungal cells.

4. Chloroplasts contain special pigments that can absorb the energy from sunlight and convert it to sugars and the chemical energy ATP, while releasing oxygen as a byproduct.

Section 2.7

1. DNA, RNA, proteins, carbohydrates, and lipids.

2. For example, proteins are used as industrial enzymes, biotherapeutics, in bioenergy production, in drug discovery, and in food science.

3. Carbohydrates are often found attached to biotherapuetic molecules and can dramatically affect their effectiveness or immunogenicity.

Section 2.8

1. DNA is complexed with proteins in the cell and ultimately condensed into chromosomes during cell division.

2. DNA is composed of four nucleotide bases (A, C, G, and T) attached to the deoxyribose sugar molecules of the sugar-phosphate backbone. The nucleotide bases pair with each other to form a double-stranded spiral or helix, with the bases in the interior of the helix. The two strands of the DNA helix run antiparallel to each other.

3. DNA is replicated in a complex process that involves many different protein enzymes. The two strands are separated and a short piece of RNA binds to one strand so that DNA polymerase may start to copy the template strand and generate a new strand. Both strands of the DNA are copied in a leading or lagging strand manner. DNA polymerase, ligase, helicase, topoisomerase, and nucleases are all involved in the process. In addition, telomerase copies the ends of the chromosomes so they do not get shorter over time.

Section 2.9

1. The cell or organism must first be lysed in the presence of quanidine (and possibly a detergent). The large debris generated may be removed by centrifugation, then the DNA in the liquid bound to a silica-based membrane. Following washing steps to remove impurities and contaminants, the purified DNA can be removed from the silica with water.

2. DNA absorbs light at 260 nanometers.

3. DNA is negatively charged and will travel in a solid support toward the cathode when subjected to an electric field, with the migration distance directly related to its size (smaller fragments migrate farther and larger fragments do not migrate as far).

Section 2.10

1. Both RNA and DNA consist of a sugar-phosphate backbone with nucleotide bases attached. The sugar found in RNA is ribose, while the sugar in DNA is deoxyribose. RNA contains a unique nucleotide base, uracil, which replaces the thymine found in DNA. RNA is single-stranded, while DNA exists as a double-stranded helix. However, RNA can bind to itself and form double-stranded regions.

2. Messenger RNA, ribosomal RNA, and transfer RNA. All are involved in protein expression.

3. RNA is synthesized through a process called transcription. It involves three main steps: initiation, elongation, and termination. During initiation, RNA polymerase and other protein helpers bind to the promoter region of the DNA and start transcription. The process in eukaryotes is much more complicated than prokaryotes, as many more protein cofactors are involved and the process is highly regulated. One strand of DNA, the template strand, is then copied into an RNA copy during elongation. The RNA polymerase uses base pairing complimentarity to create the correct sequence of bases in the RNA copy. During transcription, thymine is replaced with uracil. Termination occurs when the RNA polymerase recognizes certain features in the RNA, in conjunction with other protein factors in bacteria, and stop transcription. The RNA is then released from the transcription complex.

Section 2.11

1. For RNA purification, the first step is lysis of the target cell, tissue, or organism. This usually uses a high concentration of guanidine salt and a detergent. Following lysis, the proteins and large debris are removed by centrifugation. The cleared lysate is then processed using a silica-membrane, which will bind the nucleic acids (DNA and RNA). Following wash steps to remove the impurities and contaminants, including enzymatic digestion of the DNA, the RNA is removed from the silica membrane with water.

2. Yes, since both DNA and RNA absorb light at 260 nm. The absorbance measurement would calculate the amount of DNA and RNA present in a sample if the RNA sample was contaminated with genomic DNA from the target cells. That is why it is critical to remove the DNA from an RNA sample.

3. When total RNA is analyzed using this method, the predominant species of RNA in a total RNA sample is ribosomal RNA, of which there are two main molecules that should be visible on the gel (small and large rRNA subunits).

Section 2.12

1. 20.

2. Each amino acid has a unique structure and side chain attached. Some side chains contain a positive or negative charge at physiological pH. Others are not charged but are hydrophobic and dislike being in an aqueous environment (i.e. near water). Others are either very large or very small, and thus would influence how tightly they could be packed together in the proteins tertiary structure. All of these different characteristics influence which amino acids like to be next to each other in the final shape. They also influence what other types of proteins or molecules the protein will interact with.

3. Initiation, elongation, and termination. Initiation requires the mRNA template, the small and large ribosomal subunits, and special proteins called initiation factors. Once initiation is complete all of the components necessary for translation are present on the mRNA template. During elongation amino acids are bound to the "A" site by matching between the codon in the mRNA and the anticodon on the tRNA, then joined to the amino acid in the "P" site in the ribosome. The joined peptide is then moved to the "P" site, freeing up the "A" site for the next amino acid to come in. Translation is terminated by termination factors that cause the synthesized protein to be released and the translation complex to fall apart. Special codons indicate when translation should termination (stop codons).

4. Eukaryotic cells have only one termination factor, while prokaryotic cells have three. In addition, the features on the mRNA recognized by the initiation factors are different between eukaryotes and prokaryotes.

Section 2.13

1. Proteins can serve as enzymes, structural components, or signaling molecules in the cell.

2. The cytoskeleton is a meshwork or matrix in the cell composed of various protein components that provides structural support for the cell. It also is involved in transport and locomotion in the cell.

3. Different proteins that accomplish similar functions often contain regions of similarity that are responsible for the function. These conserved regions are often called domains. Kinase enzymes contain an ATP binding domain that is essential to their ability to phosphorylate target molecules.

Section 2.14

1. Often native proteins are expressed at very low levels, making purification and study much more difficult.

2. If you require larger amounts of protein for either study or use as a product (industrial enzyme or therapeutic), you would choose an *in vivo* expression system, while *in vitro* systems are usually used when smaller amounts of protein are required. In addition, *in vitro* systems may be used to allow for the incorporation of unusual amino acids into the protein. This might allow for different types of analysis than would be possible with *in vivo* expressed proteins. In addition, proteins that are toxic to live cells can often be expressed in cell-free *in vitro* systems.

3. Chromatography is the physical separation of a mixture of compounds by using a mobile phase and a stationary phase. It is often used for the purification of proteins based on their differential binding to various solid supports in various types of solvents and buffers. Each protein is unique and will have different binding properties to different types of matrixes (charged/polar, non-charged/nonplar, etc.).

Section 2.15

1. Size: gel electrophoresis or mass spectrometry.
   Structure: X-ray crystallography or NMR analysis.
   Identity: Western blot or ELISA.
   See text for descriptions of how these methods work.

2. Antibodies are proteins produced by the immune system that can recognize a specific molecule, such as a protein. They are utilized in Western blot analysis and ELISA assays for analyzing target proteins.

3. A functional assay is going to detect the activity of the target protein. That is, what does that protein normally do in the cell.

A Beginner's Guide to Biotechnology Unit 3

Section 3.1

1. The restriction modification system is a mechanism in bacteria to protect them from infection by viruses (bacteriophages). The bacteria express two proteins that recognize the same DNA sequence. One enzyme (restriction enzyme) can cut the DNA in this sequence, unless the sequence has been modified with a methyl group by the other enzyme in the pair (methyltransferase enzyme). The viral DNA is not modified, so gets cut by the restriction enzyme, thus inactivating it and protecting the bacteria.

2. Restriction enzymes (restriction endonucleases) recognize specific sequence in DNA, bind to that sequence, and then make a cut in both strands of the DNA at specific places. The products of a restriction endonuclease digestion are a single DNA fragment will be cut into two separate DNA fragments. The ends of the DNA fragments may be blunt (even) or sticky (uneven with overhangs).

3. BamHI: 33, 59, and 102 base pair fragments
   SmaI: 95 and 99 base pair fragments
   PstI: 103 and 1 base pair fragments

Section 3.2

1. PCR stand for the Polymersase Chain Reaction. It is a tool to amplify specific regions of target DNA so scientists can study the DNA from very minute samples.

2. The three main steps in the PCR process are: denaturation, annealing, and extension or elongation. During denaturation, the template DNA is heated to melt the dougle-stranded structure into single strands. During annealing, the temperature is lowered and the primers bind or anneal to complimentary sequence in the target DNA strands. In extension or elongation, the temperature is raised to 68-72C, and Taq DNA Polymerase recognizes the 3' ends of the primers and extends them building new DNA strand copies.

3. 1,073,741,824 copies (approximately 1 billion).

Section 3.3

1. DNA sequencing is a group of methods that can be used to determine the exact sequence of nucleotides in a fragment of DNA. Each gene has a unique DNA sequence that can be used to identify it or compare differences between different organisms.

2. Chemical sequencing determines the sequence of nucleotides in a DNA fragment using chemicals that cleave the DNA at specific sites. Chain terminator sequencing uses dideoxynucleotides to terminate a growing DNA chain during polymerization. Each of the four different dideoxynucleotides will terminate the growing DNA strand at specific sites. For either method gel electrophoresis is used to separate the different sized DNA fragments.

3. The Human Genome Project, which was started in 1990, in an international scientific collaboration to sequence the DNA in the human genome, as well as other organisms as well. This information can be used in numerous types of research to discover how organisms develop, what proteins are expressed and what are these proteins functions, and what DNA features are involved in disease.

Section 3.4

1. A plasmid is a double-stranded, circular piece of DNA found in bacteria and yeast. They are used in research as tools in biotechnology and molecular biology.

2. The different forms of plasmid DNA differ not in their size (number of base pairs), but in their structures. Some are circular and some are linear. Some are supercoiled (tightly wound), while others are relaxed.

3. Plasmid DNA is purified from bacteria by lysing (breaking open) the host bacterial cells with a very alkaline (high pH) solution. Once the cells are lysed, a solution of high salts is added to precipitate the large debris and proteins, along with the large chromosomal DNA. The solution is centrifuged to pellet the debris and the plasmid DNA is left in solution (liquid supernatant). The plasmid DNA is further purified with alcohol or capture using silica-based technologies. The purified plasmid DNA is then analyzed for purity and concentration using absorbance at 260 nanometers and the integrity determined using agarose gel electrophoresis, either with or without first digesting the plasmid with restriction enzymes.

Section 3.5

1. Plasmids contain genes that allow for resistance to various antibiotics. When transformed, bacterial cells are plated on agar containing the appropriate antibiotic, so only those bacteria that contain the plasmid will survive. This step is very important in transformation since it is a relatively inefficient process and most bacteria do not contain the DNA of interest.

2. Methods for introducing target DNA into yeast include chemical methods (lithium acetate/polyethylene glycol), particle bombardment, and protoplasts. Methods for introducing DNA into plant cells include particle bombardment, viral infection, electroporation, and the use of *Agrobacterium* to infect the plant cells.

3. Methods for introducing target DNA into mammalian cells include calcium-phosphate precipitation, cationic lipids, viral infection, and electroporation.

Section 3.6

1. Combining or joining DNA sequences together that do not normally occur together.

2. Established by the US government in 1974 in response to public concerns regarding the safety of manipulating genetic material through the use of recombinant DNA techniques, the Recombinant DNA Advisory Committee (RAC) provides guidelines for the use of recombinant DNA in the US.

3. Recombinant DNA technologies can be used for generating genetically modified organisms (GMO's), for producing proteins in bacteria, or for the production of therapeutic proteins in animal cells to name just a few.

Section 3.7

1. A recombinant protein is derived from recombinant DNA. That is, the gene from an organism is cloned into a plasmid and then expressed in a different host cell or organism.

2. Many different types of proteins are produced or used in biotechnology, but include industrial enzymes, therapeutic antibodies, growth factors, hormones, or enzymes used as research tools.

3. See text for examples. One example would be erythropoietin for treatment of anemia.

Section 3.8

1. The immune system protects organisms from invasion by foreign organisms, such as bacterial or viral pathogens. It also functions in higher organisms to protect them from abnormal cells such as tumor cells. Vertebrate's posses the most advanced and complicated immune systems.

2. An antigen is a molecule seen as foreign by the immune system. An epitope is a particular part or feature of an antigen that is recognized by individual immune B cells. An antigen can possess multiple epitopes.

3. There are 5 classes of antibodies, called IgA, IgD, IgE, IgM, and IgG. They each play a different role in an immune reaction to foreign antigens. They are also generally found in different places in the body .

Section 3.9

1. Bioremediation is the use of microorganisms (mainly bacteria), fungi, green plants, or their enzymes to return the environment to its natural state by removing toxic chemicals, oil, gasoline, heavy metals, or radioactivity

2. Phytoremediation is the use of green plants to clean up the environment, while mycoremediation is the use of fungi to clean up the environment.

3. See text for examples. One example would be the use of bacteria to decontaminate an oil spill.

Section 3.10

1. Bioenergy is the energy derived from biological sources, such as biofuels produced from biomass.

2. Biomass sources include wood and wood waste, straw, manure, miscanthus, sorghum, cassava, jatropha, rapeseed, sugar cane, sugar beets, corn and corn byproducts (like corn stover), soy and soy products such as oil, grasses, palm oil, molasses, hemp, and many others.

3. In addition to ethanol, biomass can be converted into biodiesel, electricity, syngas, methane, or other alcohols (butanol, propanol, or methanol).

Section 3.11

1. A GMO is an organism whose genetic information (DNA) has been altered in some way, usually by the addition of a gene or sequence from another organism, to hopefully confer upon the GMO a beneficial trait.

2. Transgene cassette: promoter, target gene, terminator, marker and/or resistance gene. The promoter drives expression of the target gene, while the terminator stops transcription after the target gene has been expressed. The resistance gene allows for plants containing the transgene cassette to be selected for and the marker gene allows for easy identification of plants that have incorporated the transgene cassette.

3. See text for many examples. One example would be "RoundUp Ready" corn or soy that is resistant to the herbicide RoundUp.

Section 3.12

1. DNA fingerprinting (also called DNA testing, DNA typing, and DNA profiling) are techniques used to differentiate between individuals based on their genetic makeup.

2. A short tandem repeat is a very short repeated DNA sequence (2-10 nucleotides in length. They often show variability in the number of repeats present between individuals.

3. CODIS: Combined DNA Index System. FBI-funded computer database system that contains STR DNA profiles for convicted felons.

Section 3.13

1. Mitochondrial DNA consists of a single plasmid which contains genes necessary for mitochondrial function. Because all mitochondria in a person come from their mother, mutations in the mitochondrial DNA can be traced through maternal lines. Over time small differences in the mitochondrial DNA sequences are seen and can be used to track the movement of populations around the world.

2. The International HapMap Project is a worldwide collaboration between various countries to catalog millions of SNP's in the human genome, in the hopes of linking them to various diseases, response to drug, and response to environmental compounds. They are also tracking the position of various haplotypes around the world.

3. In addition to mtDNA, SNP's and STR's present in the chromosomal DNA, including the Y chromosome, are used as well for ancestry tracking.

Section 3.14

1. Pharmacogenomics or pharmacogenetics is the study of how genes affect the reaction of the body to drugs. Different people respond differently to prescription drugs, over-the-counter medications, and things in the diet and scientists who work in this field are trying to figure out why so they can better predict how someone will respond to a drug to minimize adverse drug reactions.

2. SNP's are single nucleotide polymorphisms and are the single base differences at particular sites in a particular gene. These nucleotide changes in the DNA can in turn produce changes in the amino acid sequence of the protein, and thus potentially its function, amount, or location.

3. Cytochrome P450 enzymes are enzymes located in the liver, intestines, lungs, and skin that are responsible for metabolizing drugs or compounds taken into the body. They generally act to make them more water soluble and thus better able to be excreted. Variations in some of the CYP 450 enzymes may be responsible for how a person responds to a drug and so this class of enzymes would be screened in personalized medicine.

Section 3.15

1. The drug discovery process starts with identifying the disease that you would like to treat, then determine the target (usually a protein) responsible for that disease. Once the target is identified and verified to be the cause (or one of the causes), large numbers of diverse chemicals are reacted with the target to see if they have the desired effect (reducing its activity, increasing its activity, binding to the correct receptor, etc). This process is called primary screening. Any compounds that are identified in this initial test, are then confirmed and tested to see what other targets they might inadvertently effect which could cause side effects. Medicinal chemists try to make the potential drug compound better (more efficacious and less toxic) and at some point the potential drug compounds are tested in animal models for safety and efficacy as well. Only once sufficient safety has been shown in animals can human clinical trials begin to show safety and efficacy in healthy humans and patients with the disease. Following sufficient testing and data generation the FDA may approve the drug for sale.

2. The US Food and Drug Administration play's a crucial role in new drug development in the US. It over see's the entire process to verify that any potential drugs introduced into the market are efficacious against the advertised medical condition and are not overly toxic or unsafe.

3. Biotechnology allows for vaccines to be developed to single bacterial or viral proteins or mixtures of purified proteins, or purified nucleic acid (usually DNA), such that live, attenuated whole virus or bacteria, or even dead virus or bacteria, are not required to generate an immune response. They may be safer and more specific.

Printed in Great Britain
by Amazon